Emergency Incident Risk Management

Emergency Incident Risk Management

A Safety & Health Perspective

Jonathan D. Kipp

Murrey E. Loflin

JOHN WILEY & SONS, INC.
New York • Chichester • Weinheim • Brisbane • Singapore • Toronto

Copyright © 1996 by John Wiley & Sons, Inc.

Originally published as ISBN 0-442-1926-2

No part of this publication may be reproduced, stored in a retrieval system, or transmitted in any form or by any means, electronic, mechanical, photocopying, recording, scanning or otherwise, except as permitted under Sections 107 and 108 of the 1976 United States Copyright Act, without either the prior written permission of the Publisher, or authorization through payment of the appropriate per-copy fee to the Copyright Clearance Center, 222 Rosewood Drive, Danvers, MA 01923, (978) 750-8400, fax (978) 750-4744. Requests to the Publisher for permission should be addressed to the Permissions Department, John Wiley & Sons, Inc., 605 Third Avenue, New York, NY 10158-0012. (212) 850-6011, fax (212) 850-6008, E-mail: PERMREQ@WILEY.COM

**For ordering and customer service, call 1-800-CALL-WILEY.
ISBN # 0-471-28663-X**

10 9 8 7 6 5 4 3 2

Library of Congress Cataloging-in-Publication Data

Kipp, Jonathan D.
    Emergency incident risk management : a safety and health perspective / Jonathan D. Kipp and Murrey E. Loflin.
       p. cm.
    Includes bibliographical references and index.
    ISBN 0-442-01926-2 (hc)
    1. Emergency management–Safety measures.   2. Risk management.
    3. Fire fighters.   I. Loflin, Murrey E.   II. Title.
HV551.2.K56   1996
363.3'7'0684—dc20   96-1108
                                                                           CIP

Contents

Preface xi
Acknowledgments xv

PART 1. ADMINISTRATION AND ORGANIZATION 1

1. Overview 3
   Scope of Problem 5
   Learning Lessons 6
   Risk Management 7
   Risk Management Program 8
   History of Health and Safety in Emergency Services 10
   How to Use This Book 11

2. Introduction to Risk Management 13
   Introduction 15
   Benefits of Risk Management 16
   Roles and Responsibilities 19
   Support and Active Participation 23
   Program Compliance 24
   Conclusion 26

3. Accident, Injury, and Illness Data 29
   Introduction 31
   Data Collection/Reporting Processes 31
   Why Keep Records? 35
   Confidentiality 39
   Conclusion 39

4. Law, Codes, and Standards 41
   Introduction 43
   Occupational Safety and Health Administration 44
   Federal Mandates 49
   State Laws 50
   Consensus Standards 50
   Standard Operating Procedures (SOPs) 56
   Influence and Effect of Laws, Codes, and Standards 56
   Periodic Review and Revision Process 57
   Conclusion 57

PART 2. COMPREHENSIVE RISK MANAGEMENT PLAN 59

5. The Management of Risk 61
   The Process of Managing Risk 63
   Choices 64
   Goals and Objectives 65
   Risk Retention 66
   Administrative Risk Management Versus Emergency Incident Risk Management 66
   Classic Risk Management Model 67
   Conclusion 68

6. Risk Identification
   Introduction 71
   Risk Identification Methods 73
   Sources of Information 74
   Recordings of Findings 80
   Conclusion 80

7. Risk Evaluation 81
   Introduction 83
   Evaluation Measures 83
   Frequency and Severity Considered Together 88
   Conclusion 91

8. Establishing Priorities 93
   Introduction 95
   Analysis Considerations 97
   Balancing the Analysis Factors 103
   Establishing Priorities 103
   Conclusion 105

9. Risk Control 109

Contents　　　　　　　　　　　　　　　　　　　　　　　　　　　　　　　　vii

　　　　Introduction　109
　　　　Risk Control Techniques　109
　　　　Risk Assumption and Risk Financing　119
　　　　Conclusion　119

10.　Program Monitoring　123
　　　　Introduction　125
　　　　Program Effectiveness　125
　　　　Frequency of Monitoring　127
　　　　Who Conducts the Evaluation?　128
　　　　Evaluation Methodology　130
　　　　Results of Evaluation　132
　　　　Conclusion　133

11.　Training of Personnel　135
　　　　Introduction　137
　　　　Training—A Vital Component of Pre-Emergency Risk Management　138
　　　　Training as a Risk Control Technique　140
　　　　Accident Prevention and Training　141
　　　　Live Training Evolutions　142
　　　　Mandated Training　144
　　　　NFPA 1500　145
　　　　Risk Management and Training　153
　　　　Conclusion　156

PART 3.　EMERGENCY INCIDENT RISK MANAGEMENT　157

12.　Pre-Emergency Risk Management　159
　　　　Introduction　161
　　　　Written Risk Management Plan　162
　　　　Written Safety and Health Program　163
　　　　Health and Safety Officer Function　164
　　　　Toolbox　168
　　　　Conclusion　175

13.　Principles of Emergency Incident Risk Management　177
　　　　Evaluation of Conditions　179
　　　　Pre-Emergency Risk Management　180
　　　　Pre-Incident Planning Program　180
　　　　Target Hazard Classification　183
　　　　Completing the Pre-Incident Plan　183
　　　　Pre-Incident Preparation　187
　　　　Conclusion　195

14. Incident Safety Officer   197
    Introduction   199
    Responsibility and Authority   200
    Response Criteria   201
    Incident Management System   202
    Emergency Authority   202
    Duties and Functions   203
    Incident Scene Monitoring   204
    Forecasting   209
    Post-Incident Analysis   211
    Conclusion   211

15. Personnel Accountability   213
    Introduction   215
    Philosophy   217
    Reasons for A Personnel Accountability System   217
    Concept of the Personnel Accountability System   232
    Pass Devices   232
    The Players   233
    Standard Components of the Personnel Accountability System   234
    System Features   236
    Conclusion   243

16. Incident Management System   245
    Introduction   247
    Objectives of the Incident Management System   248
    The Inception of Incident Management   250
    Incorporating Risk Management into the Incident Management System   252
    Toolbox for Evaluating Incident Risk   256
    Conclusion   259

17. Post-Incident Analysis   261
    Introduction   263
    Benefits and Components of the Post-Incident Analysis   264
    Safety and Health Issues   265
    Interfacing with the Incident Safety Officer   270
    Health and Safety Officer's Responsibility   271
    The Occupational Safety and Health Committee's Responsibilities   273
    Paradigm Shift for Health and Safety   274
    Conclusion   274

PART 4. INTEGRATION   275

18. Making It Happen   277

Introduction 279
Benefits of Effective Risk Management 280
Process Versus Event 281
Safety Is A Value 281
Tips for Making It Happen 282
Sample Risk Management Plan 282
The Future 284

Appendix A   Common Risks and General Control Measures   285

Appendix B   Virginia Beach Fire Department Risk Management Plan   291

Appendix C   Sources of Additional Information   299

Select Bibliography 303
Index 305

Preface

Risk management is a complex topic. Reams upon reams have been published in an effort to help those who practice risk management improve their skills. There are many different disciplines that impact the risk management process, including safety, loss control, finance, security, and insurance, to name only a few.

Risk management in a career context also has several different connotations. For example, an investment risk manager has responsibilities dramatically different from those of an insurance risk manager. However, the basis upon which risk management decisions are made remains similar.

Our focus in this work is the health and safety of the professionals who respond to emergencies and help others. They may spend a great deal of time preparing for the short time they will spend controlling any given emergency incident, yet that short time is when the need to use effective risk management skills is most important. Too many responders continue to be injured and die in the line of duty.

We have attempted to present the foundation upon which effective risk management decisions appropriate for those who must make them quickly on a daily basis can be structured. The chapters are arranged similarly to an incident command chart; as your risk management process expands, so too do the concepts presented in the book.

To many, attempting to manage risks may be similar to trying to stretch a rope around a fogbank: nebulous at best, impossible at worst. That may be the reality in many situations. However, with the basics in hand, sound decisions will be easier to make.

We are not professional risk managers. Rather, we have been, and continue to be, in positions in which we recognize the need for improved decision making regarding personnel health and safety. We believe that this work can help to remedy some of those deficiencies.

Others have made the methodology we discuss work, and we know that our readers can too.

Jonathan D. Kipp
Londonderry, NH

Murrey E. Loflin,
Virginia Beach, VA

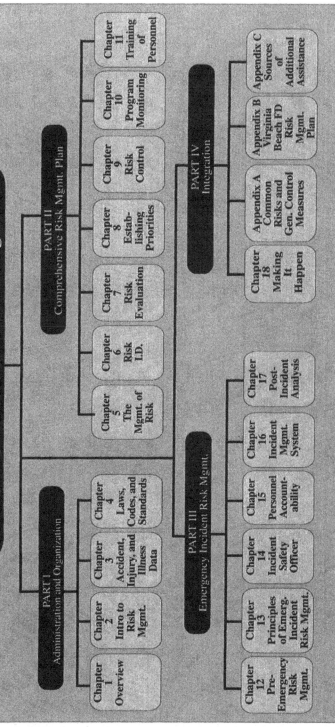

Acknowledgments

I am indebted to the many people who helped to make this book a reality. It is impossible for me to thank each one of them individually, but several in particular deserve recognition.

My wife, Martha, and our children, Ellen and Robert. They tolerated more late nights, weekends without me, and my preoccupation with "the book" than I will ever be able to make up for.

My Mom and Dad. I owe them both a great deal. Dad put his priorities in this order: first the family, then the fire service, then dentistry.

Murrey Loflin, my coauthor. His expertise combined with his relaxed disposition served as an excellent counterbalance for my lack of each.

Paul Genovese, Chief Executive Officer of Compensation Funds of New Hampshire, my employer. Paul encourages all staff member to get involved in projects that will reflect positively on themselves, and therefore the organization. I hope that occurs in this instance.

Malisa Denney of Brandford, Massachusetts. The graphics in the book are entirely her work, and she deserves enormous credit not only for her expertise and creativity, but also for her patience and understanding in having to deal with me.

Many professionals in the fire service whom I have come to know and admire over the years, including John Rines, formerly of the Durham-UNH Fire Department; Dennis Parker, former chief of the Collegeville, PA Fire Company; Alan Brunacini, chief of the Phoenix Fire Deparment; Don Bliss, New Hampshire State Fire Marshall; and Bruce Teele and Stephen Foley of the NFPA. I'm sure there are others, and for those of you I left out, I apologize.

To all of you, thanks.

—Jonathan Kipp

Although I accomplished the literal task of developing, researching, and writing this book, I could not have completed this work without the assistance of my family and friends.

I would like to thank my mother and father, Mabel and Paul Loflin, Sr. They instilled in me the ethic of striving to succeed. Mom and Dad, I thank you for your support, direction, and advice. To my sisters and brother—JoAnne Gilchrist, Sarah Borcherding, and Paul Loflin, Jr.—thank-you for your love, support, and encouragement. I must say a special thank-you to JoAnne for her constant guidance and input during this project. I would like to thank my mother- and father-in-law, Marilyn and Don McAdams, for their encouragement during this project.

I joined the fire service in June 1979 as a fire fighter with the City of Beckley (WV) Fire Department. Prior to that I had spent four wonderful years at Marshall University, working towards an undergraduate degree and gaining experience as a fire fighter. The late John H. McCulloch, then Mayor of the City of Beckley, hired me upon graduation from Marshall. The three years I spent with the Beckley Fire Department served as the basis for my development as a fire fighter. I would like to thank the officers, personnel, and the civilian staff of the Beckley Fire Department for sharing their experience and knowledge. I have to say a special thank-you to Wayne Westfall, John Thomas, David Underwood, and Elizabeth Settle for their support.

Since February 16, 1983, I have been fortunate enough to be a member of the Virginia Beach Fire Department. The officers, personnel, and the staff represent the finest fire service organization in the country. Fire Chief Harry E. Diezel has had the true vision to develop a fire department occupational safety and health program, one of the first in the country. Chief Diezel gave me my start as a safety officer. I have had the pleasure of working for Deputy Chief James W. Carter for the past eight years; he has had the responsibility of ensuring the success of the safety and health program. I also thank Robert Esenberg, Risk Manager for the City of Virginia Beach, for his guidance with the risk management process. I would like to thank Battalion Chief Donna P. Brehm, for sharing her fire service knowledge and experiences with me as well as supporting my role as safety officer. Over the past several years I have had an opportunity to work with Battalion Chief Michael W. Wade on many safety and health issues. Thank you, Mike, for your assistance and guidance.

Other fire service personnel that I would like to thank for their continual guidance and support are Fire Chief Alan V. Brunacini of the Phoenix Fire Department, Fire Protection Engineer Kevin M. Roche of the Phoenix Fire Department, NFPA Senior Fire Protection Specialist Stephen N. Foley, NFPA Senior Fire Protection Specialist Bruce W. Teele, and especially all the members of the original NFPA Fire Service Occupational Safety and Health Technical Committee.

I must thank my partner and friend Jonathan D. Kipp. I have truly enjoyed working with Jonathan and gaining a better understanding of the risk management

ACKNOWLEDGMENTS

process as it applies to the fire service. I hope this is the start of a continuous partnership.

Finally, to my family. To Maureen, my dear wife and the mother of our children, for her countless hours of reading manuscripts. This book would not have been possible without her support, guidance, and insight. She is my inspiration and the driving force behind me. Thank-you, Maureen. To my son Reilly Stewart. I thank you for your input. Though only five years old at this time, he loves to be involved and to participate in my projects. Reilly, you are always welcome to participate, and never lose that keen imagination. My daughter Haley Royar was born during the course of writing and is a welcome addition to the family. What a true bundle of love, joy, and happiness. Thank-you, all.

—Murrey Loflin

Emergency Incident Risk Management

PART 1 | Administration and Organization

PROFILE
Chapter 1: Overview

MAJOR GOAL:

To lay the foundation for the importance of effective risk management

KEY POINTS:

- Understand that the record of firefighter accidents, occupational injuries, and occupational illnesses has been poor.

- Define and understand the concept of risk management from a health and safety perspective.

- Define and understand the concept of a risk management program.

- Review the history of occupational health and safety in emergency services.

- Outline the contents of the book, and explain how to use it as a reference guide.

Emergency Incident Risk Management

PART I
Administration and Organization

**Chapter 1
Overview**

Chapter 1

Overview

SCOPE OF PROBLEM

Newspaper headline: *3 Firefighters Killed, Scores Injured!* Was it written in 1925 or in 1995? If the headline read *New Model T Sells for $290*, would you be able to tell? Why is it that for emergency service providers very little has changed in those 70 years?

Headlines relative to injuries among emergency service providers remain all too common in our daily newspapers. Firefighters and other providers have an unenviable accident, injury, and fatality record. The easiest way for a firefighter to be named in the news is to get killed or injured. There is a deep-seated problem with any system in which there is perceived positive recognition (being named in print) for behavior that has a negative outcome (being injured).

Annual studies of firefighter fatalities have been conducted by the National Fire Protection Association (NFPA) since 1974 (see the graph in Figure 1.2). The number of on-duty firefighter deaths in the United States peaked at 171 in 1978 and averaged 133 from 1977 through 1988. Only in 1989 did we begin to see a steady decline in this statistic, as deaths went from 117 in that year to 75 in 1992 and 73 in 1993. Unfortunately, however, the number rose back up to 100 in 1994. The average for the six-year period 1989–1994 is 97.

Injuries to firefighters are far more difficult to track than are fatalities, but many individual attempts are made to collect accurate data. Best estimates are that 100,000 firefighters are injured each year in America. This number does not include industrial firefighters or emergency medical response personnel, so the record can only be worse than is indicated by the numerical estimate.

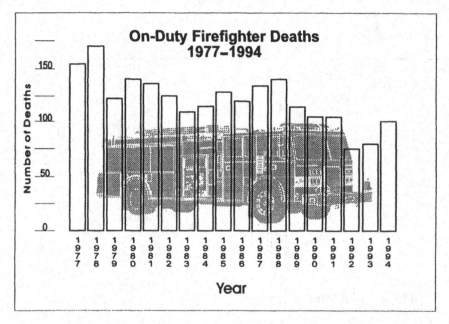

Figure 1.2. On-duty firefighter deaths in America, as reported by the NFPA, 1977–1994.

LEARNING LESSONS

It is unfortunate, but all too often the life of a firefighter is the price the fire service pays to learn lessons. In both training exercises and actual emergency incidents, we continue to relearn the same tragic lessons.

During training exercises, many factors have historically led to deaths and injuries, a primary one being lack of compliance with National Fire Protection Association (NFPA) Standard 1403, *Standard on Live Fire Training Evolutions in Structures*. Key components of this standard include an effective command structure, use of appropriate personal protective equipment, adequate preparation and supervision, and a personnel accountability system.

Following actual emergencies, investigations have documented the dangers associated with fighting fires in buildings that have wood truss roofs, for example. Other incidents have indicated the importance of having a personnel accountability system in place and operating, and also of being able to utilize an appropriate fire ground communications system.

Emergency responders suffer too many deaths, injuries, and illnesses in the course of their employment, and a logical question is: "Why do we need to keep relearning these same lessons, at such an exorbitant cost?" That, unfortunately, is a difficult question to answer. Every year, responders die, are injured, or become ill as a result of conditions that frequently can and should have been identified and controlled.

However, it is not the purpose of this book to point out an obvious problem. The purpose, rather, is to outline the theory and practical applications of sound risk management strategies that can help provide solutions. Effective, proactive management of the risks that confront personnel who respond to emergencies is a concept that is just beginning to develop. Other works have addressed the issue of controlling liability risks for fire departments, but this book focuses primarily on the health, safety, and survival of the people who risk their lives to save the lives and property of others.

Risk management as a thought process or decision-making tool can and should apply to all aspects of any organization. Though in this book we shall concentrate on personnel health and safety, readers are encouraged to apply these same risk management principles in other disciplines as well.

The concepts presented in the book are not limited by organizational boundaries, mission statements, or geographical borders. Firefighters and other professionals who are called upon to respond to emergencies face the same risks regardless of employer or location. Whether employed by a city or a town, a major corporation or the government, whether located in a metropolitan area or a wilderness area, in the United States, Europe, or Asia, theirs is a dangerous business. Risk management is exempt from artificial boundaries, and can and should have a positive impact on any organization.

RISK MANAGEMENT

One textbook definition of the risk management process reads as follows:

> **Risk Management Process:** System for treating pure risk; identification and analysis of exposures, selection of appropriate risk management techniques to handle exposures, implementation of chosen techniques, and monitoring of the results.[1]

This definition is straightforward, relatively clear, and concise, yet meaningless to many unless used in an appropriate context.

For our purposes, we shall consider risk management to be a sort of toolbox in which are kept the various tools needed for ensuring the health and safety of

---

[1] McIntyre, W. S., and J. P. Gibson. *Glossary of Insurance and Risk Management Terms*. Dallas, TX: International Risk Management Institute, Inc., Fourth Edition, 1983, 1986, p. 167.

personnel. Included in this toolbox are items such as Standard Operating Procedures (SOPs), a process or system for managing incidents that includes a personnel accountability system, and a health and safety program. Carpenters carry their tools to the job in a toolbox, and remove the most appropriate tool when it is needed. The same is true for effective risk management. The risk management toolbox provides a variety of resources that can be used to positively impact the health and safety of personnel.

RISK MANAGEMENT PROGRAM

Now that you understand better what risk management is, we can move on to a discussion of a risk management *program*. Very simply, such a program is a systematic method of applying the basic principles of the management of risk. When these principles are worked into the fabric of an organization and made an integral part of the decision-making process, the risk management program will be in place.

NFPA 1500,[2] *Standard on Fire Department Occupational Safety and Health Program,* in Section 2-2, calls for the establishment of a risk management program, and reads as follows:

2-2   Risk Management Plan

2-2.1 The fire department shall adopt an official written risk management plan that addresses all fire department policies and procedures.

2-2.2 The risk management plan shall cover administration, facilities, training, vehicle operations, protective clothing and equipment, operations at emergency incidents, operations at nonemergency incidents, and other related activities.

2-2.3* The risk management plan shall include at least the following components:

    (a). Risk Identification: Potential problems;

    (b). Risk Evaluation: Likelihood of occurrence of a given problem and severity of its consequences;

    (c). Risk Control Techniques: Solutions for elimination or mitigation of potential problems; implementation of best solution;

    (d). Risk Management Monitoring: Evaluation of effectiveness of risk control techniques.

---

[2]Reprinted with permission from NFPA 1500, *Standard on Fire Department Occupational Safety and Health Program,* Copyright © 1992, National Fire Protection Association, Quincy, MA 02269. This reprinted material is not the complete and official position of the National Fire Protection Association, on the referenced subject which is represented only by the standard in its entirety.

Although it may contain few words, Section 2-2 requires a great deal in terms of actual compliance procedures. To comply with NFPA 1500, a department must adopt a comprehensive program that addresses all of its operations. There are some who believe that this should have been a requirement in the initial (1987) edition of NFPA 1500, because it is almost impossible to outline solutions for potential problems that have not been identified in terms of operations.

Risk management is further emphasized in Chapter 6 of NFPA 1500, where risk management strategies for emergency operations are outlined. The premise is basic, and steeped in common sense: "Risk a lot to save a lot, risk nothing to save nothing." This premise does make sense, but it poses a dilemma for safety professionals or risk managers. Section 6-2.1.1 in Chapter 6 declares that an employee will risk his or her life to save a life. Few experts in other fields would create a safety or risk management program that included a section that made it acceptable for employees to risk their lives! However, fighting fire or responding to other emergencies is not a typical industry.

In 1983, a group of dedicated volunteers began work on what would become the 1987 edition of NFPA 1500, *Standard on Fire Department Occupational Safety and Health Program*. Led by Alan Brunacini, chief of the Phoenix, AZ, fire department, the group refused to accept the idea that employee accidents, injuries, and illnesses had to be accepted as just another part of the job. They composed the first edition of the standard and purposely designed it to be an umbrella document, under which additional, companion documents would be drafted in the years ahead.

The fact that the 1987 edition of NFPA 1500 included numerous groundbreaking requirements for firefighter health and safety was difficult to accept for many who would be affected by the document. There were some claims that the requirements would result in the demise of the American fire service as it had been known up to that time. However, there were many others who applauded the effort, but who questioned the reasoning upon which the requirements were based.

When consensus standards or laws are developed, the organizations that will be most affected expect and deserve to understand the logic behind the particular requirements. The answers to questions regarding the basis of the requirements originated from a blend of history, experience, common sense, and judgment, exercised by both the members of the technical committee and the public through the NFPA's standards-making process. After all, after thinking about the situation, who can question the effectiveness of requiring the use of self-contained breathing apparatus (SCBA) during an interior structural fire attack?

An initial requirement for the implementation of a risk management plan in the 1987 edition of NFPA 1500 would have allowed individual departments to determine their own needs and formulate their compliance plans accordingly. This is an easy observation to make, given that hindsight is always accurate. However, the requirement is now included in the 1992 edition, and rather than being viewed as just one more exercise in paperwork, fire department administrators ought to consider it an opportunity to better define the needs of their entity and to plan accordingly.

History of Health and Safety in Emergency Services

There are several textbook definitions of *insanity*. A non-textbook one that is easy to understand is: "doing the same thing repeatedly, but expecting different results." In the past, how many times have firefighters been exposed to toxic products of combustion without self-contained breathing apparatus, and been expected not to suffer smoke inhalation? Such behavior and such an expectation qualify as insane.

Risk management has only a brief history in emergency services. Unlike other professions or industries, responding to emergencies has inherent risks. Over the years, it has been expected that response personnel would accept risk as part of their job, and not hesitate to risk, or even sacrifice, a life, in order to save someone else's life or property. That expectation is changing.

Organized firefighting did not begin in America until 1679, when the city of Boston, MA, formed the first paid fire department. Earlier, fire wardens were used in New York City, beginning in 1648, to inspect chimneys for flaws that could lead to fires. In 1658, New York City had what was known as a "rattle watch." This consisted of individuals who would patrol the streets and, upon discovering a fire, would sound an alarm by shaking wooden rattles. Ben Franklin created the first volunteer fire department in 1736, in Philadelphia.

Initially, firefighters used bucket brigades and hand-drawn fire apparatus. The organizations that provided these services were considered by many to be social organizations, and the competition between "companies" could be strong. Many times, the focus in responding was on beating the neighboring company to the scene of an incident rather than on controlling the emergency.

The competition led to the taking of chances by the members. Injuries, although not well documented, were frequent. As a result, some methods to protect the members were initiated. Rudimentary helmets were provided, as well as rubber boots and a raincoat. The primary purpose of these articles was to protect the responders from the elements, as well as from the ever-present water and soot that accompany a working fire.

Eventually, firefighters began to conduct interior structural firefighting operations. The motivation behind this was a desire for efficiency and a need to comply with the demands of insurance companies. In the early stages of fire insurance, companies were known to reward well those fire brigades that limited damage in an insured building. This is one reason why fire marks that indicated an insured building were so valuable to have and display; a fire in a marked building was likely to be attacked very aggressively.

However, with these operations came increased risk. Again, the only line of defense against the risk was personal protective equipment. Helmets, coats, boots, and gloves began to evolve into more-specialized pieces of equipment. With the advent of self-contained breathing apparatus (SCBA) following World War II, the respiratory system became another part of the body that was protected.

Even with these efforts, the record of accidents, injuries, illnesses, and fatalities among responders was dismal. Traditions of machismo, "leather lungs," and self-sacrifice to save people and property were formidable. Many have described this situation as 200 years of tradition unimpeded by progress.

The tradition sustained an embarrassing trend of member injuries and fatalities. Firefighters also continued to die from heart attacks and other stress-related maladies. Personal protective equipment, although improved, did not represent a panacea; it still had to be properly selected, sized, and, most importantly, used—and, in too many cases, was not.

In the 1970s, things began to change. Rather than being an individual effort on the part of one entity, the safety and health process became a more collective interest. Thanks in part to the American space program, turnout gear became more sophisticated and consequently more effective. Mandatory use of self-contained breathing apparatus became more common. Standards addressing personal protective equipment, apparatus, and personnel safety came under development, and were eventually adopted. Incident command systems for managing various types of emergencies (primarily wildland and high-rise incidents) began to be utilized.

Although they were not described this way at the time, these developments were the beginning of risk management for emergency services. Individual components were developed based on their own merits, but few people saw this as a risk management program for the organization.

HOW TO USE THIS BOOK

This is a reference book. It is not intended to be read once and set aside. Rather, it is organized to provide information and guidelines that will help the reader to design and implement an effective risk management program. There are five major parts to the book, as outlined below. Each segment stands on its own and is appropriate for individual reference.

Part 1   *Administration and Organization*
    The basics of risk management from an administrative standpoint,
        including a discussion of applicable laws, codes, and standards
Part 2   *Comprehensive Risk Management Plan*
    Establishment of a framework for a comprehensive, pre-emergency
        risk management plan for the organization

Part 3 *Emergency Incident Risk Management*
  The nuts and bolts of assessing and controlling the risks on the scene of an emergency
Part 4 *Integration*
  A review of the key components of the book, and how they all tie together
Appendix A Common Risks and General Control Measures
Appendix B Virginia Beach Fire Department Risk Management Plan
Appendix C Sources for Additional Assistance

Each chapter begins with a profile that summarizes the major goals of that chapter, lists the key points that are covered, and suggests methods for implementation by the reader. Note that action may be required after consideration of the points listed. Some of the concepts included are thought provoking, and will challenge users of the work to ask themselves and their organizations some questions. Many of the questions will have difficult, or possibly disturbing, answers. Ultimately, however, action taken after an evaluation of those answers will help to improve the health and safety of firefighters and other emergency responders.

PROFILE

Chapter 2: Introduction to Risk Management

MAJOR GOAL:

To provide a comprehensive introduction to the process of managing risk

KEY POINTS:

- Understand the differences between non-emergency and emergency incident risk management.
- List some of the many benefits to effectively managing risks, including:
  - financial benefits
  - improved efficiency
  - improved health and safety
  - greater likelihood of compliance with applicable laws, codes and standards
- Understand that every member of the organization, both individually and as part of the following groups, has responsibility for the success of the risk management program:
  - the "organization"
  - top management
  - middle management
  - the employees
  - health and safety officer
  - risk manager
- Point out that although the roles may be filled by the same individual, there are differences in the duties of the Health and Safety Officer (HSO), and the Incident Safety Officer (ISO).
- Understand that the Health and Safety Officer typically oversees the administrative risk management program. The Incident Safety Officer functions as the emergency scene risk manager.

Emergency Incident Risk Management

PART I
Administration and Organization

- **Chapter 1** Overview
- **Chapter 2** Intro to Risk Mgmt.

Chapter *2*

Introduction to Risk Management

INTRODUCTION

*R*isk *management* may mean many different things to many different people. People in the business world are paid a great deal of money to manage risks. Most people are not, and for them risk management is probably not something to which they pay much conscious attention.

In reality, however, we all manage risks every day. Backing the family car out of the driveway requires an analysis of the risks involved and the taking of actions to manage those risks appropriately. If there is oncoming traffic, we wait, because we do not want to *risk* having a collision.

There has to be a balance, however, in taking risks. Taking an inordinate number of risks or allowing known risks to go uncontrolled can, and usually will, result in severe damage or injury. On the other hand, taking no risks results in inactivity, a paralysis driven by indecision and fear. We lead life by effectively balancing these two extremes of risk and safety.

To better understand risk management, we must first define *risk*. From there, we can move on to a discussion of what risk management is and the benefits that are realized by having an effective risk management program in place. You will learn that there are several definitions and assumptions about risk management because its applicability varies with the people and the situations that it impacts.

The definition of *risk* in NFPA 1500 is as follows: A risk is a measure of the probability and severity of adverse effects. These adverse effects result from an exposure to a hazard.[1]

Classic risk management, when viewed from a "corporate" perspective, has basic elements that have been used for many years. Historically, risk management has been viewed as an insurance purchasing and loss prevention function. Only recently have risk managers been asked to help identify risks to their organizations, provide guidance on appropriate control measures, and handle the financial risks presented. These same elements are important for a provider of emergency services, because the basic principles of running a business are the same. However, this is only the administrative side of the business.

Each member of an emergency services organization must also contend with the risks associated with *providing* those services. This is the characteristic that most dramatically illustrates the difference between emergency service providers and other entities.

For people who respond to the emergencies of others, the risks are numerous and significant. After all, firefighting, treating victims of medical emergencies, and carrying out rescues in confined spaces, to name only a few such situations, are inherently dangerous. People do not call for assistance when things are going well. They wait until their own risks are uncontrolled before they call for help. Help arrives in the form of highly trained personnel who have the proper tools, equipment, and apparatus to control and mitigate the incident.

However, the mere arrival of the experts does not suddenly cause the risks to become controlled. Those experts, whether they be firefighters, emergency medical technicians, paramedics, or hazardous-materials technicians, are required to manage the risks that their customers decided they couldn't handle themselves.

Why are the experts required to manage the risks? The consequences of not doing so can be as serious as life and death, or as simple as ensuring that the paint on the apparatus doesn't get scratched. Regardless of the stakes, a risk that goes uncontrolled or unmanaged creates a problem for the members of that organization. The people who effectively manage their risks reap many benefits for their efforts.

BENEFITS OF RISK MANAGEMENT

Effective risk management can yield many benefits, some more tangible than others. For example, making money or, at a minimum, not losing money, is a generally accepted goal of organizations, and effective risk management will

---

[1]Reprinted with permission from NFPA 1500, *Standard on Fire Department Occupational Safety and Health Program,* Copyright © 1992, National Fire Protection Association, Quincy, MA 02269. This reprinted material is not the complete and official position of the National Fire Protection Association, on the referenced subject which is represented only by the standard in its entirety.

Introduction to Risk Management

help to ensure that that occurs. More importantly, effective risk management can also help to ensure the continued health, safety, and well-being of the organization's personnel. The "cradle-to-grave security" concept is gaining more widespread recognition; people who choose the profession of responding to emergencies should be able to enjoy a productive, healthy retirement after a successful career. Retirement should not be a sentence served in hospitals and rehabilitation centers.

Several benefits are outlined in this section. Every reader will have his or her own perspective, so we make no judgment about the order of priority of these benefits. To some, financial benefits may be most important, whereas to others health and safety may be paramount. Any benefit that can be realized is important, and our list is presented in that context.

Financial Benefits

A quality risk management program saves money. For example:

Fewer accidents and injuries should occur.
The ones that do occur will, it is hoped, be less serious.
Recovery after a serious loss will be quicker and less disruptive.
Insurers will have greater confidence in their policyholders' ability to limit losses and thus claims, so premiums may be lower.
With fewer accidents and injuries, efficiency is improved.

Improved Efficiency

Utilization of a well-organized risk management effort can pay dividends to the organization. Beyond preventing accidents and injuries, it can also lead to better efficiency. This efficiency is then translated into an improved decision-making environment, which allows the organization to function as effectively, and therefore as safely, as possible.

Risk management programs are dynamic in nature, and require continuous review and revision. The benefit of this is that problems can usually be identified early enough that they do not become too cumbersome, time consuming, or expensive. A problem left unchecked only grows, so early and effective intervention is important.

The advent of the use of Incident Management Systems brought a high degree of control and structure to the incident scene. This has had a dramatic positive impact on the health and safety of responders. If that same principle is applied to all other duties, the same results can be expected.

This point is actually clearer when viewed in reverse. Chaos is confusion, and confusion leads to the taking of unnecessary risks. The taking of unnecessary risks leads to losses, whether they be personnel losses (injuries and illnesses) or property losses (vehicle accidents and fires, for example). The bottom line is that the more organized any effort is, the more likely it is to be completed efficiently, effectively, and safely.

Safety and Health

The primary focus of NFPA 1500 is the safety and health of personnel. One of the first responsibilities outlined in the document is the development of a risk management plan that addresses all functions of the organization. The philosophy of the members of the NFPA Fire Service Occupational Safety and Health Technical Committee is that an effective risk management plan can have a positive impact on firefighter health and safety.

That philosophy is based upon the evidence presented by companies that have enjoyed those benefits for years. The DuPont company has long utilized an intense, effective program that does not tolerate losses. The impact has been dramatic, and it is now a core belief within the organization that accidents are not inevitable.

Although it is called by several different names, such a process represents risk management. Risks are identified, and control measures are instituted to ensure that losses won't occur. The process works.

Compliance

A successful risk management plan will allow a more orderly, comprehensive review and understanding of applicable laws, codes, regulations, and standards. With that understanding comes an opportunity for a higher degree of compliance.

Because one of the components of the risk management process is the identification of appropriate control measures for the hazards that create risks, standards that impact those controls should also be identified. Frequently, standards that address health and safety outline the required end result (controlled risks), so compliance becomes a secondary benefit to outlining the safest way to perform a task.

Summary of Benefits

All of these benefits can be realized because a risk management program provides the foundation for an organized approach for the identification and control of risks. A quality program addresses all risks confronting an organization, many

Introduction to Risk Management

of which are easily identified and understood. In those cases, control measures are usually straightforward. However, there are other risks that are not so easily identified, and that is where a comprehensive program with a systematic approach becomes important so that no risks are overlooked.

ROLES AND RESPONSIBILITIES

Who are the key players in a risk management program, and what are their roles and responsibilities? A poll of so-called "experts" would yield a typical assortment of answers to these questions, such as the following.

First-line supervisors are the link between management and workers. They understand the work flow, and can balance the needs of all parties while still getting the job done.

Top management is made up of the decision makers, who are the only ones who can truly set policy. Without top management, and their commitment, the program will go nowhere.

The employees are the backbone of the organization. Without their acceptance and understanding, the risk management program, or any program, for that matter, is doomed to fail. Top managers and first-line supervisors can hand down all the rules, policies, and directives they want, but unless the employees "buy in," it will be a fruitless exercise.

The department's health and safety officer or risk manager is the one who is paid to handle this risk management, and, if the risk management program fails, isn't it his or her responsibility?

All of these answers have merit, the last one the least, but none of them is totally correct. A thorough examination and evaluation of each of the affected groups or individuals (the organization, top management, supervisors, employees, health and safety officers, and risk managers), their roles, and their responsibilities under a risk management program will provide insight into what makes a program tick.

The Organization

Every member of an organization has a responsibility within the guidelines of the risk management program. The responsibilities vary by position, but no one is exempt from the duty of maintaining a safe, healthful, and essentially risk-free workplace. This is frequently viewed as the "organization's" role in the risk management program, the assumption being that the organization can somehow control the risks posed by doing business. However, an organization is nothing without people, so risk management is truly a responsibility of individuals, not of an organizational structure. Therefore, when discussing the responsibilities of the "organization," we are actually addressing all members of that organization.

Top Management

Call them what you will—top management, the upper echelon, the office of the chief—these are the people in charge of making sure the mission of the organization gets carried out, that the job gets done, that it gets done safely and efficiently, and that money is not lost in the process.

From a risk management standpoint, these are also the people who are responsible for ensuring that an effective program is in place and operating, and that others are carrying out their responsibilities under the program.

Probably the most effective role of the top manager(s) is to demonstrate genuine support and commitment for the risk management program. This will make it clear to the members of the organization that risk management is important, that they will be held accountable for their actions or inactions under the program, and that the organization can and will benefit from using the program as a tool for controlling losses. A sample "Statement of Safety Policy" is shown in Figure 2.2.

If top management is effective in transmitting this message, it will become clear that member safety is a value within the organization, not simply a priority. Why is member safety an organizational value rather than a priority? Priorities are assigned, and can and do change. Something that is a high priority today may not be tomorrow. Values do not change.

However, the health and welfare of members can never be anything but the top priority. For that reason, it needs to be a value. It cannot be dismissed during some arbitrary reassignment of priorities. By being part of the organization's value system, its status is safeguarded. When this is achieved, when safety is truly an organizational value, all operations, all decisions, will include that safety component, but typically on an automatic, subconscious level. No separate, distinct "safety" decisions will be required.

Values are sometimes described as the underpinnings of an organization. They are timeless, and define what the organization stands for. Through time, through countless changes, through other factors that require the organization to adapt, core values remain. Daily practices are driven by the invariant core values.

It is not easy to make member health and safety a part of the organizational values system. To do so requires many things, not the least of which is time and commitment. Top management must make it clearly understood at every opportunity that nothing will be allowed to compromise safety. Whether during a budget discussion or at the scene of a major emergency, actions and decisions speak louder

STATEMENT OF SAFETY POLICY

The City of Douglastown values the health, welfare, and safety of every employee and intends to provide a safe and healthful workplace. The intent of this policy is to reduce the frequency and severity of accidents, injuries, and illnesses, for they cause untold suffering and financial loss to our employees and their families. In pledging its full support for the safety process, the mayor and city council recognize certain responsibilities, as follows:

1. That prevention of accidents and protection of all resources are guiding principles.

2. That all operational decisions affecting safety must receive the same consideration as those affecting efficiency, or quality.

3. That safe working conditions and methods are of prime importance and take precedence over shortcuts or other "quick fixes."

4. That employees will function during all operations, both emergency and non-emergency, in the safest manner possible.

5. That the Douglastown Fire Deparment will comply with all safety laws and regulations.

6. That feedback will be welcomed from all employees.

7. That all employees will follow all safety procedures, take no unnecessary chances, use all appropriate safety equipment, and make safety an integral part of their lives.

As an employee of the Douglastown Fire Department you have a responsibility to yourself, your family, your co-workers, and the community to understand and follow our safety process. We must be alert in detecting and taking steps to remedy potentially hazardous conditions. Above all, we must exercise concern for others to help ensure everyone's safety, well-being, and productivity.

Your efforts will make the difference!

_____   _____
Fire Chief Signature   Date

Figure 2.2. A sample Statement of Safety Policy.

than words. One misstep and credibility is lost. If top management implements a policy but then does not follow it, that fact is clear to others.

Middle Management

Middle managers (those individuals with the authority, but not the complete responsibility, for leading others) hold the most unenviable positions in any organization. They serve as the link, or bridge, between top management, on the one side, and employees on the other.

Top managers are always demanding more of middle managers but providing them with fewer resources. Workers have difficulty understanding why their supervisor doesn't sympathize with their position and do more to improve their lot. With these two groups (top management and employees) pulling at them from different directions, middle managers are often stretched to the limit.

Among the multitude of their responsibilities, two of the most important are providing effective leadership and providing effective discipline. Leadership demonstrates and promotes positive behavior. In the context of the risk management program, this leads to safer, more productive employment.

As we mentioned for top management, appropriate behavior cannot merely be described to and expected of others; it must be demonstrated. Middle managers have almost constant contact with other employees, and are frequently looked to for leadership, so it is critical for them to set and follow the standard. Many texts have described and analyzed the difference between managing and leading, but one of the most important is setting the standard for others and following it yourself.

On the other hand, discipline is required when an employee violates any of the organization's policies. For many, initiating discipline is the least favorite, least rewarding part of being a supervisor. However, when it is carried out effectively, discipline can have a positive impact on the organization.

Disciplinary procedures usually follow some sort of progression. A common system starts with a verbal warning, and goes on to a written warning, suspension, and, ultimately, termination. Other factors that may affect an organization's disciplinary process are applicable laws, union contracts, and past practice. Regardless of the system used, the discipline employed should fit the infraction. The progression outlined can be invoked at any step, but judgment is important to ensure fairness and consistency.

Consistency is critical. The star employee and the troublemaker must be disciplined in the same manner when they violate the same policy. Otherwise, the organization and the people who represent it will quickly lose credibility. This, again, impacts the value system. All members of the organization are expected to operate under the same set of rules, and they should also be expected to receive the same discipline, when it is required.

A common misconception is that violators of safety and health policies should fall under a disciplinary process that is separate from the process used for other infractions. In fact, it is more appropriate to allow the organization's personnel policies and procedures to address discipline regardless of the transgression. This will help to ensure uniformity, and also provides for clearer understanding by all parties on what discipline to expect when it is required.

The Employees

Support, active participation in, and compliance with a program can be viewed as the prime responsibilities of all members of the organization—*support* because no program can succeed without belief in it by those who *actively participate* in it, and *compliance* because workers need and expect a valid set of guidelines under which to operate.

SUPPORT AND ACTIVE PARTICIPATION

How is program support achieved, measured, and maintained? Each of these is a key component, and none is easy to accomplish. The starting point, however, is a well-designed, quality program. It is far easier for people to buy in to something that is well constructed than something that is slipshod, poorly executed, and ineffective.

Scholars have long said that credibility can only be earned and that, once lost, it is extremely difficult to recover. Program support is indeed similar. If those who must operate under the program understand it, recognize its goals, and believe in and agree with those goals, they will support and participate in it. However, they must be educated and trained in all aspects of the program to gain that much-needed understanding. Otherwise, whether they are required to follow the program or not, that program is doomed to fail.

Workers who are actively concerned about themselves, their family, friends, and co-workers are typically responsive to efforts on the part of the employer to maximize health and safety. Active concern involves not only caring about self and others, but actively demonstrating it. Rather than reporting a hazard to a supervisor, an actively concerned employee takes the necessary steps to control and alleviate the hazard—for example, wiping up the puddle of water versus reporting it to the maintenance department!

Active concern carries over to an employee's personal life as well. It is unreasonable for employees to expect and demand the employer to go to great lengths to ensure their health and well-being on the job when they take no responsibility for these factors on their own. Risk management is everyone's responsibility both on and off the job, and it must be a team effort. Each party should expect the other to do all that is possible; with this partnership it can become a "win–win" situation for all.

Health maintenance and physical fitness are excellent examples of responsibilities that span the entire organization. Few need to argue the benefits of being healthy and physically fit, but many argue the motivation behind an employer's requiring these qualities in his or her employees. What is the true agenda in such a situation? A new policy that all employees must undergo a medical evaluation is rarely seen as a benefit of employment. Rather, it is viewed as an effort on the part of the employer to "weed out" unwanted employees. Imagine the frustration of an administrator who plans to provide medical evaluations for a group of employees, obtains all necessary approvals and funding, but is rebuffed by the very employees he or she is trying to help!

Such failure doesn't have to be the case. A well-planned and executed policy will have had input from all affected employee groups. The communication and education required to achieve acceptance and active participation will have been conducted prior to program implementation. That way, all questions and misconceptions will have been addressed.

This process is just that—a process. Education and communication must be continuous. The fact that a program has been implemented does not guarantee that it will always have the support and active participation of those it affects. Employees change, and the assumption that "everyone has bought in" is dangerous! The group in place initially may have accepted the program, but, because turnover occurs, the communication and education must be continuous. If not, aspects of the program may later be questioned, and it may be necessary to start the process from the beginning again.

Program Compliance

Employees provide the support for a program and actively participate in it. However, their participation needs to be guided in some fashion to maintain order. Just as society passes laws to prevent anarchy, organizations have rules, regulations, and guidelines to set the parameters for organizational behavior. Workers both expect and deserve a valid set of work rules.

Once valid work rules are established, they need to be consistently observed, applied, and enforced. A rule without enforcement is merely a suggestion. Therefore, the organization needs to have an enforcement mechanism in place, and those charged with executing this enforcement must clearly understand the process, be able to explain it to others, and be able to use it.

Health and Safety Officer

On paper, the top manager, such as the chief of department, will be listed as the individual responsible for the organization's risk management program. In reality, the organization's health and safety officer (HSO), should the position exist, is likely to carry out the duties.

This position is still relatively new in many organizations, and became better defined with the publication of the 1987 edition of NFPA 1501, *Fire Department Safety Officer*. This document, renumbered *NFPA 1521* in the 1992 edition, outlines the qualifications for, and duties and responsibilities of, the position.

More recently, a distinction has been drawn between a department health and safety officer and an incident safety officer. The health and safety officer will likely have responsibility for many of the administrative aspects of the organization's risk management and safety programs. Questions about OSHA compliance, recordkeeping, and the organization's accident experience will all be directed to the HSO.

Typically of more importance to the employees on the front line is the incident safety officer (ISO). This individual is in a position to watch over the health and well-being of all members during operations at an emergency incident. It should be comforting for a firefighter to know that during an interior structural attack there is somebody outside, with direct access to the incident commander, whose responsibility it is to monitor actions and/or conditions that may jeopardize the health and safety of members. If any hazards are observed, the incident safety officer can immediately notify the incident commander so that appropriate action can be taken. The duties and responsibilities of the incident safety officer are further discussed in Chapter 14.

Frequently, the health and safety officer and the incident safety officer will be the same individual. Regardless of whether there is one person or a full staff, anyone serving in the capacity of a safety officer needs a variety of qualifications, including the following.

- Working knowledge of, and experience with, department operations, tactics, and deployment policies;
- Ability to stay current on health, safety, medical, and fitness issues that may affect the organization;
- Ability to communicate effectively with all personnel, at all levels of the organization;
- Familiarity with applicable health and safety laws, codes, standards, and regulations.

From an administrative risk management standpoint, some of the things that the health and safety officer may get involved with include the following.

- Risk identification and classification;
- Recommendation of appropriate risk control measures;

If so directed, the formulation and implementation of safety-and health-related policies and procedures;
Safety- and health-related administration and recordkeeping;
Employee counseling following accidents, exposures, and near-misses;
Accident investigations;
Communication with others outside the organization, such as other safety officers, risk managers, medical professionals, and insurance professionals.

This list is certainly not all-inclusive, and the entries will vary based on, among other things, the type and size of the organization. A large metropolitan fire department will likely have several health and safety officers, whereas a volunteer department serving a small rural community may incorporate the safety officer responsibilities with other functions, such as those of a training officer.

In organizations in which the emergency response personnel are actually employed in other disciplines, but respond during the incident, the safety officer may be a full-time safety professional. Frequently, with a plant fire brigade, the employees will respond when summoned, and the organization's on-site safety director, or the equivalent, will also respond and fulfill the responsibilities of the incident safety officer.

Risk Manager

Some organizations are fortunate to have a position dedicated solely to the risk management function. Large corporations have had risk managers for years, but more often we now see the title *municipal risk manager* in the public sector. This individual will typically have oversight over the risk management programs of several departments, emergency services among them. However, many of the functions can be centralized, which will relieve the department's administration from some of the risk management tasks.

Most frequently, the risk manager will handle relations with outside agencies such as insurance companies and/or agents, and be responsible for handling the insurance needs of various departments. In addition, this individual is familiar with the overall risk management process, and can serve as a valuable resource for guidance and information.

CONCLUSION

Risk management, when implemented and practiced effectively, can provide many benefits for the members of the organization. However, nothing comes without effort. Everybody in the organization has responsibilities under the risk management program, and the contributions of each member are important.

Chapter 3 discusses the role of information in the risk management process. The responsibilities outlined in this chapter cannot be properly carried without quality data that then become useful information. Chapter 3 addresses the sources of those data, and ways they can and should be used.

PROFILE

Chapter 3: Accident, Injury, and Illness Data

MAJOR GOAL:

To understand the importance of quality data for the risk management effort, where to find it, and how to use it

KEY POINTS:

- Understand that data are useful only when turned into information.

- Be familiar with the various data reporting/collection efforts.

- Understand the differences between the data reporting/collection efforts that are mandatory, and those that are voluntary.

- Recognize that there are many reasons for collecting/reporting data. Evaluate each one to ensure the benefits are worth the effort.
  - Data analysis
  - Mandates
  - Legal liability
  - Benchmarking
  - Medical / insurance

- Know what is, and is not, confidential, and how to handle both.

- Recognize that "garbage in is garbage out." Maintain and report only quality data.

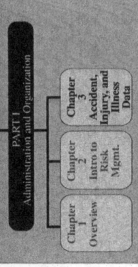

Chapter 3

Accident, Injury, and Illness Data

INTRODUCTION

Good data are critical for making good decisions. This is especially true in the risk management process. So much of our planning is based on history that it becomes vital to maintain, and have access to, good data, which then must be turned into good information. In this chapter, we shall address some of the various types of data and resulting information, their availability, the strengths and limitations of each, and some means of using the information to help with the formulation and adoption of an effective risk management plan.

However, complete data are not readily available to us yet. Of the numerous data collection efforts that we shall address in this chapter, none captures the entire injury and illness experience of emergency response personnel. Some try, but due to the diversity of department types, varied jurisdictional regulatory requirements for reporting, perceived value (or danger) in reporting data, and lack of interest, some information never leaves the location where it is generated.

DATA COLLECTION/REPORTING PROCESSES

There are several data collection, reporting, and analysis processes in existence. Some are mandatory and some are voluntary. They each involve their own purposes for gathering the data, and different methods as well. Each entity requires an understanding of the local, regional, state, and federal filing requirements. From there, any of the several voluntary data collection systems may be chosen. Regardless of which of these is chosen, each participant should be assured that he or she can retrieve usable, relevant data that can be converted to information. Data that are reported but are not retrievable in a timely, usable format are useless. Some of the data collection/reporting efforts/systems are addressed in the following sections.

Insurance Companies

Insurance is a valuable tool in the risk management arsenal, and an in-depth discussion of the topic is included in Chapter 9. Regardless of whether they purchase insurance or are self-insured, most, if not all, organizations will deal with the insurance issue. As such, there are data required for operating an insurance program, and those same data are available for use and analysis.

From an employee accident, injury, or illness standpoint, the workers' compensation program administrator or insurance carrier will be the best source of information. In order to qualify for benefits, employees are required to submit some form of documentation about their injury or illness. This information is then used to determine whether or not to provide benefits and, if so, at what level.

A summary of all reports filed by a particular entity will be useful for identifying trends. This information should be available for the asking. In some cases, however, *persistence* may be the key word, but insurance companies will provide the data when pressed.

In addition, many insurance companies provide some form of loss control or loss prevention assistance, and the loss data can serve as a good initial point for discussion with a loss prevention representative. This same type of statistical information is usually available for other lines of insurance coverage as well. For example, a record of apparatus accidents or property damage claims can be requested from the appropriate insurance carrier.

International Association of Fire Fighters

The International Association of Fire Fighters (IAFF) has produced an annual firefighter death and injury study since 1960. This report includes statistics from departments that have a membership in the IAFF from the United States and Canada.

Report data generated by the IAFF include statistics on lost work time, disability retirements resulting from accidents and injuries, emergency medical service (EMS) injuries, and physical fitness injuries. The data are more thorough than those collected by the National Fire Protection Association, but are limited to career departments with an IAFF affiliation.

Local Jurisdictions

Regardless of any mandatory or voluntary reporting mechanisms, all emergency service organizations should maintain information relating to the death, injury, or illness of department personnel. They should also adopt procedures for proper and thorough documentation of accidents or incidents affecting department equipment, vehicles, and facilities.

Both NFPA 1500, *Standard on Fire Department Occupational Safety and Health Program*, and NFPA 1521, *Standard on Fire Department Safety Officer*, have sections dealing with data collection and maintenance. Though not mandatory requirements, these consensus standards recognize the need for maintaining quality data.

Frequently, the organization's insurance carrier or administrator will dictate what information and documentation is required for their purposes. From an employee safety and health standpoint, this information can also provide valuable insight that can be used to positively impact the safety and risk reduction efforts for personnel.

With the health problems associated with exposures to hazardous materials or communicable disease, high-quality recordkeeping and documentation are critical. Adverse health effects or illnesses associated with exposure to chemicals and communicable diseases may not appear for years. Inadequate or nonexistent documentation of an exposure can seriously impact insurance and medical coverage for members.

Agencies responsible for medical and health coverage for personnel have very strict guidelines for reporting and documenting accidents, injuries, and illnesses. Failure to comply with these guidelines can be costly to the department and to the individual member, not only financially, but also in terms of emotional and physical anguish.

National Fire Protection Association

In 1974, the National Fire Protection Association (NFPA) began an annual study that collected data on firefighter fatalities in the United States. Until 1995, this process was conducted under a contract with the United States Fire Administration (USFA) that included a more extensive analysis of patterns and trends in specific parts of the firefighter fatality problem.

Each year the NFPA also surveys a sample of fire departments in the United States. The sample is carefully selected to represent the various-sized populations served by fire departments. The results of the survey are used to project, with a high degree of confidence, the *national* firefighter injury experience. However, it is not an actual survey of all departments.

Both the firefighter fatality survey and firefighter injury survey are published by the NFPA in their periodical *Fire Journal*. In addition, reprints of past articles are available from the NFPA.

Occupational Safety and Health Administration (OSHA)

The Occupational Safety and Health Administration (OSHA) is an agency of the Federal Department of Labor. It has five major areas of responsibility, as follows:

1. To promulgate, modify, and revoke safety and health standards;
2. To conduct inspections and investigations and to issue citations, including proposed penalties;
3. To require employers to keep records of safety and health data;
4. To petition the courts to restrain imminent danger situations; and
5. To approve or reject state plans for programs under the Act.

These requirements affect public fire departments in 23 states and 2 territories that have their own Occupational Safety and Health Administration (OSHA) programs. The remaining states come under the guidelines of federal OSHA, which specifically excludes coverage for state and local government employees.

In order to ensure that employers record and maintain injury data relating to employees, OSHA established recordkeeping requirements. Every employer subject to OSHA must maintain specific information on what is called the "OSHA 200 Log." This log must be posted each year from February 1 through March 1 in a conspicuous space for all personnel to view.

The recordkeeping requirements call for the employer to record information about every occupational death, nonfatal occupational illness, and nonfatal occupational injury. To be considered a "recordable" occupational injury, one or more of the following must have occurred to the injured employee: loss of consciousness, restriction of work or motion as a result of the injury, transfer to another job as a result of the injury, or receipt of medical treatment other than first aid.

Other information required by the OSHA 200 Log includes the injured employee's name, job title, and work assignment, type of injury or illness, type and extent of medical treatment, and length of lost work time, if any. The purposes behind these recordkeeping requirements are to enforce the OSHAct, to study the causes and prevention of occupational accidents and illnesses, and to maintain useful statistics on occupational safety and health.

Statistical surveys and the establishment of methods used to acquire accident and injury data are the purview of the Bureau of Labor Statistics (BLS). OSHA delegates to the Commissioner of the BLS its authority for the development and maintenance of a program to collect, compile, and analyze occupational safety and health statistics.

Public Safety Officers Benefit Program (PSOB)

The Public Safety Officers Benefit Program is administered by the Federal Department of Justice. Signed into law in 1970, the program provides a death benefit to the survivors of public safety officers, including firefighters, whether career or volunteer, who are killed in the line of duty.

There are specific reporting requirements and forms to be used to initiate a claim, and most of the information is used to verify that the death occurred in the line of duty. The PSOB maintains little in the way of a data base that is useful for identifying measures that can be taken to prevent future fatalities. The NFPA fatality studies are far more useful for those purposes.

United States Fire Administration

The National Fire Incident Reporting System (NFIRS) is a standardized format used for reporting fire loss to the United States Fire Administration. Though not all fire departments participate in this process, most or all states collect fire loss data with this process.

One component of this process, the Fire Service Casualty Form, is for reporting firefighter injuries and fatalities. The data recorded are for incident-related injuries and fatalities only. The definition of "incident-related" is the period from the time the alarm sounds to the time the apparatus is placed back in service. If a firefighter injury or fatality occurs in any other function it is not recorded on this form.

As with the other data collection efforts, this base of data is not complete either. Participation in the program is voluntary, and there are several reporting forms required for the various components of the incident (fire loss data, firefighter casualty, civilian casualty, etc.). This makes it impossible to predict how many fire service casualties go unreported to this system. Further, the ones that are reported will exclude injuries or fatalities that occur during operations such as training and maintenance.

WHY KEEP RECORDS?

In the previous section we discussed the sources for much of the data that are available for review. But why keep the information? From a risk management perspective, there are numerous reasons for maintaining quality data, which range from legal requirements to planning, to budgeting. Examples of some of the reasons follow.

Data about the number, cost, and nature of worker injuries and apparatus accidents will help with the design of control strategies.

The maintenance of a health data base will allow for the tracking of illnesses and exposures for each worker, thus aiding in the determination of job-relatedness later.

Workers' compensation experience impacts future costs for the insurance, and that can dramatically affect the budget.

Benchmarking will provide the framework for comparison between organizations.

These and other reasons for collecting, reviewing, and understanding data are further outlined in the following sections.

Data Analysis

If we can identify what has happened in the past, and why, we may be able to prevent it from happening again. This is one of the most important uses of data available to any organization. Various types of information will be useful in this planning and decision-making process. At a minimum, the following should be included on any report so that it will be available to be analyzed later:

Date and time the incident occurred
Employee's name and other appropriate information such as address, social security number, and telephone number
Current position assignment or job title
Name of the employee's immediate supervisor
Date and time the supervisor was notified
Exact street location of the incident
Location and type of activity the employee was engaged in at the time of the incident, such as:
    at an incident scene (include type of incident)
    while working at the station
    while responding to or returning from an incident
    while performing training
A brief but thorough description of how the incident occurred
Type of injury and body part(s) affected, or description of loss or damage to equipment, vehicle, or facility
If appropriate, the cost of repair or replacement
Actions taken to prevent a recurrence, or recommendation(s) to prevent the accident or injury from recurring

This information, when taken as a whole, will provide an excellent summary of past experience. Based on the conclusions drawn, actions can be taken to alter any trends. Rarely is there a single cause of an accident. Usually a combination of factors converge and an accident occurs. Some of the more common areas that have historically lead to accidents or near misses follow:

Employee action or inaction
Lack of training, or inadequate training
Poor condition of, or lack of use of, appropriate personal protective equipment
Lack of use of, or inadequate, personnel accountability system
Inadequate personal health maintenance

Whether through a specific accident investigation or a review of data, operations and/or performance that should change to positively affect the safety and welfare of the organization's personnel and resources can be identified. Next, an action plan should be developed to address these changes. A quality action plan includes information on the changes or revisions that need to take place, who is responsible, the dates the changes will be made and when they will become effective, and any other required details.

An interesting perspective on data analysis is the challenge of identifying today what will be important later. In order for data to be credible, there has to be a body of data. That usually means that they have been collected for a period of time, and any anomalies have had time to be absorbed by what has become a sizable body of data.

What should be collected now for analysis later? For example, there is an ongoing debate about the effects of staffing levels and employee accidents and injuries. When the debate began, there was no credible bank of data from which to draw conclusions because nobody had collected and maintained specific information about staffing levels progressively throughout incidents. If that information were available, much of the controversy could have been avoided.

Mandates

Some data are required to be maintained. Examples are the OSHA recordkeeping requirements for those entities subject to OSHA, and any state requirements for exposures to communicable diseases. Although we believe that there is no value in collecting data that cannot be later used productively, the threat of fines or other sanctions is enough to recommend that mandated records be kept.

Legal Liability

The best defense against a liability claim is quality documentation. Regardless of the severity of an accident, injury, or incident, collect and record as much information as possible. Be certain that the following information is recorded, in addition to the information listed earlier under *Data Analysis*:

Names and addresses of witnesses
Statements made by witnesses, bystanders, and any injured parties
Site conditions
Weather conditions
Vehicle positions, if appropriate
Anything else that may impact the case

Depending on the nature of the accident, some of this information may be gathered by others, such as law enforcement personnel. The important point is that the information be available for later retrieval in the event of litigation.

Memories fade with time, so witnesses recalled years later may have forgotten details that may be important in a case. If their statements are recorded at the time of the incident, the risk of losing a critical piece of information is reduced.

Benchmarking

Benchmarking has become a buzzword of the '90s. There are many who say that its time has come and passed, whereas others claim that it provides a legitimate basis for evaluating a risk management program.

Benchmarking was born out of the Total Quality Management (TQM) philosophy. In its purist sense, it allows for the comparison of processes and results between two or more groups, with one group being held superior; that is the benchmark. Because risk management is frequently a subjective discipline, benchmarking may not always apply. When it comes to data analysis, however, benchmarking may provide a consistent basis for comparing one organization's accident and illness records with those of another. In effect, it can become a quantifiable measure of success or failure.

Whether benchmarking will endure in the lexicon remains to be seen. However, it is likely that justification for resources needed for the risk management program will continue to be required, so some means for determining and measuring progress will be developed.

Medical/Insurance

In addition to the mandated recordkeeping discussed above, insurance companies and other medical providers also usually require certain records to be maintained. Health histories, records of treatments and medications, diagnoses, and results of medical evaluations are examples. Some of these data will typically be maintained by the provider, and some by the employer.

Because of patient confidentiality issues, individual health information is not, and should not be, available to others in the organization. However, health data for a *population* of employees can yield valuable information about that population. Just as an evaluation of accident and injury data can help with the identification of trends, and possibly suggest appropriate control measures, the same is true of these aggregate health data.

CONFIDENTIALITY

There are strict guidelines regarding confidentiality associated with medical and health records, and it is important for the organization to understand and observe them. It is also important to recognize the differences in the guidelines that apply to patients treated by members of the organization, and those that apply to the members themselves.

Members need to be trained in all aspects of patient confidentiality. Policies addressing the required paperwork and the issue of who will have access to it and where it will ultimately be stored need to be in place. In addition, as part of their training, members need to understand the guidelines associated with information gathered during patient care, and their responsibility for handling that information.

Medical records for members must also be treated in a confidential manner. Medical records include such information as records of job-related accidents, illnesses, and injuries, doctor's reports, and fitness evaluations. Based on United States federal confidentiality statutes, the Americans with Disabilities Act, and the Canadian Freedom of Information and Protection of Privacy Act, issues such as who will maintain control of the records and where they will be stored must be addressed. Whether the medical records relate to patients or members, legal counsel should be consulted for assistance. The penalties for violation of the statutes can be severe.

CONCLUSION

Recordkeeping and documentation are critical elements of a department's risk management process. This responsibility rests primarily on the health and safety officer, who must ensure that procedures are followed, that documentation is completed in an accurate and timely manner, and that all the bases are covered.

ADMINISTRATION AND ORGANIZATION

The information generated as a result of data collection and analysis can help the organization in several ways. Most importantly, future accidents, injuries, and illnesses can hopefully be prevented. Trends can be identified early and reversed. Measures can be instituted to control what may otherwise may become an out-of-control situation that will adversely affect the organization and its members. Lastly, potential liability can be limited.

PROFILE
Chapter 4: Laws, Codes, and Standards

MAJOR GOAL:

To understand the laws, codes, and standards that impact the operations of the organization and help ensure compliance with these requirements

KEY POINTS:

- Identify applicable laws, codes, and standards that are mandated for compliance.

- Attend training courses and seminars to become educated and familiar with the operations of the organizations that develop laws, codes, and standards.

- Develop a network that provides avenues for information and updates, and allows input for revision.

- Incorporate compliance with laws, codes, and standards into the standard operating procedures of the organization.

- Develop a system that tracks the compliance requirements for each mandate for all members of the organization.

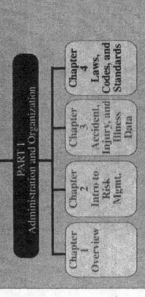

Chapter *4*

Laws, Codes, and Standards

INTRODUCTION

The industrialization and technological advances at the turn of the century greatly impacted the workplace in the United States. Instead of a business, company, or industry focusing on productivity at any cost, safety mandates and workers' rights began to emerge as parallel issues. However, many employers focused strictly on productivity and profit, with little or no regard to the health and welfare of their workers. Safeguards for equipment and machinery were in most cases nonexistent. If an employee was injured, disabled, or killed, the employee was simply replaced, with little or no compensation to the employee or relatives. Most employers were exonerated of any criminal or civil penalties because no laws relating to compensation or employer liability existed at the time.

With improvements and advances in technology, employers were forced to confront the issue of employees' rights, safety, and health. The development of labor laws for children makes this evident. These laws protected children's rights relating to employment requirements, type of work environment, safety, and hours worked. Safety and health issues such as workplace fire prevention and protection, safeguards for equipment and machinery, and medical treatment for occupational injuries were addressed by the employer. Yet another issue that surfaced was workers' compensation for occupational injuries and accidents. Employers had to provide means for compensating employees injured on the job, regardless of the severity of the injury or who was at fault.

In this chapter, we shall examine several of the laws, codes, and standards that directly impact an organization. It is imperative to explore this issue from a legal and liability standpoint. We must assume that all organizations are operating within the confines of applicable laws, standards, and regulations. The organization must have the documentation to support the way it operates during an emergency incident. These are specific guidelines that we must follow in order to achieve our standard outcome.

OCCUPATIONAL SAFETY AND HEALTH ADMINISTRATION

On December 29, 1970, the Occupational Safety and Health Act of 1970 (Public Law No. 91-596) was signed into law by President Richard M. Nixon. The OSHAct had an effective date of April 28, 1971. Coauthored by Senator Harrison A. Williams (D-N.J.) and Congressman William Steiger (R-Wis.), the Occupational Safety and Health Act is referred to as the Williams–Steiger Act or the Williams–Steiger Occupational Safety and Health Act. Prior to the passage of the OSHAct, legislation for general industry was intermittent and grossly inadequate. Most states had little or no safety and health legislation, inadequate safety and health standards, insufficient staffing, and inadequate enforcement procedures. Budgets were poorly funded. In turn, the annual occupational fatality rate was greater than 14,000 employees killed, and 2.2 million suffered work-related injuries annually in the late 1960s. Prior to 1970, the federal legislation for occupational safety and health was erratic. Efforts were directed more toward specific interests than the widespread coverage designated in the OSHAct.

When the OSHAct of 1970 was enacted, each state was given the opportunity to develop its own safety and health plan. Under the Occupational Safety and Health Act of 1970, federal OSHA has no provision for direct enforcement to ensure that state and local governments comply with safety and health standards, such as the OSHA bloodborne pathogens standard for public employees. OSHA law does permit other methods to be utilized in order to maximize the protection of public employees' safety and health.

Twenty-three states and two territories have established and maintain an effective and comprehensive occupational safety and health program for public employees. These state plans must meet or exceed the requirements of federal OSHA. OSHA gives each state 6 months from the publication date of any new federal final standard to adopt a similar state standard. All fire departments, whether state, county, or municipal, in any of the 23 states or 2 territories that have an OSHA plan agreement in effect have the protection of the minimal acceptable safety and health standards mandated by federal OSHA.

The 23 states and 2 territories that have state OSHA plans that cover public employees are:

| Alaska | Kentucky | North Carolina | Virginia |
| Arizona | Maryland | Oregon | Virgin Islands |

Laws, Codes, and Standards 45

California	Michigan	Puerto Rico	Washington
Connecticut	Minnesota	South Carolina	Wyoming
Hawaii	Nevada	Tennessee	
Indiana	New Mexico	Utah	
Iowa	New York	Vermont	

Regardless of whether a department is mandated to comply with the requirements under the OSHAct or not, it is good practice to do so for the protection and welfare of personnel.

With more emphasis being placed on health and safety, plus the advent of NFPA 1500, fire and EMS agencies are now more aware of pertinent laws and standards. Laws and standards are a critical and significant part of the operations of any emergency services organization. Administrators must remain knowledgeable on and compliant with current safety and health regulations. Managers responsible for the delivery of customer services must have a thorough understanding of the laws, standards, and regulations that govern the safety and health of members relating to response and actions at the emergency scene. The following laws and standards provide a basis for a solid safety and health program:

29 CFR 1910.120, *Hazardous Waste Operations and Emergency Response*
29 CFR 1910.146, *Permit-Required Confined Space*
29 CFR 1910.156, *Industrial Fire Brigades*
29 CFR 1910.1030, *Bloodborne Pathogens*

29 CFR 1910.120

29 CFR 1910.120, *Hazardous Waste Operations and Emergency Response,* provides specific requirements for agencies that respond to hazardous materials incidents. Because of training, education, resources, and availability, the fire service predominately responds to these incidents. With the passage of this law, EMS providers are also required to institute particular risk management components within their written plans.

First, the law requires that members working at an incident site consider the following risks:

Exposures exceeding the permissible exposure limits and published exposure levels
Concentrations that are immediately dangerous to life or health (IDLH)
Potential eye irritation sources
Explosion sensitivity and flammability ranges
Oxygen deficiency

Second, the law requires a proactive health maintenance program that monitors the health and welfare of team members. Third, it requires the use of an effective incident management system that can incorporate the response of additional agencies and personnel outside the needs of the respective department. Fourth, it requires the use of personal protective equipment as outlined in the appropriate NFPA Standards. The training and competency levels are clearly defined in an effort to prevent personnel from performing job tasks beyond their skill level.

29 CFR 1910.146

The Occupational Safety and Health Administration (OSHA) has initiated safety requirements, which include a permit system for entry into confined spaces, under 29 CFR 1910.146, *Permit-Required Confined Space.* These confined spaces pose specific hazards for entrants because their design hampers efforts to protect entrants from significant hazards, such as toxic, explosive, or asphyxiating atmospheres. This standard provides extensive guidance for employers for protecting their personnel who work in permit-required spaces.

The *Permit-Required Confined Space* standard specifically addresses permit space hazards. Most OSHA standards provide limited protection and, based on a review by OSHA, employees are not sufficiently protected from atmospheric, mechanical, or other hazards. OSHA must ensure that employers continue to monitor, test, and communicate at workplaces that contain permit-required confined spaces. Employers must have a written program in place that addresses the organization's needs relating to this standard.

The section that affects fire and other emergency service personnel is Section (k), *Rescue and Emergency Services.* There are specific guidelines for emergency services that would allow employees to enter permit spaces to perform rescue services. The following requirements apply.

> Employer shall ensure that each member of the rescue service is provided with and is trained to properly use the personal protective equipment and rescue equipment used to make rescues.
> Each member shall be properly trained to perform the assigned rescue duties and shall receive the necessary training under paragraph (g) of the standard.
> Members must make at least one practice permit space rescue every 12 months from actual permit spaces or from representative permit spaces.
> Each member must be trained in basic first aid and cardiopulmonary resuscitation (CPR). At least one member that has current certification in first aid and CPR shall be available.
> A host employer that arranges to have personnel other than their employees perform permit space rescue must inform the rescue service of the hazards they may confront at the host employer's facility.

Laws, Codes, and Standards 47

The host employer must provide rescue personnel with access to all permit spaces so that these personnel can develop appropriate rescue plans.
Non-entry rescue retrieval systems or methods must be used whenever an authorized entrant enters a permit space.
Material safety data sheets (MSDS) must be available as required.

There are approximately 239,000 confined spaces in workplaces throughout the country. In the past, multiple deaths have occurred in confined-space rescues when the would-be rescuers also became victims. Emergency services personnel have a poor safety record in trying to remove victims from confined spaces. Approximately 80–90% of such fatalities could be avoided if employers and employees followed the requirements of 1910.146. OSHA has determined that asphyxiation is the leading cause of death in confined spaces.

Emergency service agencies must determine if they are capable of providing personnel and resources to perform rescue safely from confined spaces. They must all identify additional resources to assist in the event of an emergency of this type. Planning is the key to success in this type of rescue. Identify whether your community has the resources to handle these types of incidents safely and effectively. Knowledge of the standard, training, and education are issues that must be addressed before attempting to respond to and handle such an emergency.

29 CFR 1910.156

29 CFR 1910.156, *Industrial Fire Brigades* standard, provides requirements for training and education, protective clothing and equipment, and respiratory protection for firefighters. It also requires that fire departments provide a statement or written policy that validates or provides information on: the permanence of the fire department; the organizational structure; type, amount, and frequency of training for members; the expected number of members in the fire department; and the functions the members are to provide at the work site.

Obviously, these mandates were written before the development of NFPA 1500, *Standard on Fire Department Occupational Safety and Health Program.* The requirements of the *Industrial Fire Brigade* standard still mandate that organizations ensure that personnel be properly trained to accredited levels before they are allowed to participate in structural firefighting activities. This also requires that personnel have the proper protective clothing and equipment to conduct these operations.

Training and Education

The training and education requirements mandate that personnel receive the proper instruction that is commensurate with the duties and functions they are expected to perform. This training and education will be conducted prior to participation in emergency operations. All fire department personnel must be provided training at least annually; for those who conduct interior structural firefighting, the training must be provided at least quarterly.

On December 1, 1980, this regulation went into effect for the public and private fire departments in the 25 states and territories with state OSHA plans. Compliance criteria that had an immediate effect on the fire service were mandates for self-contained breathing apparatus and protective clothing.

Self-Contained Breathing Apparatus (SCBA)

SCBA used for structural firefighting was required to have a 30-minute rating and automatic audible alarm when the air supply was depleted to a specified level, and to be of positive-pressure type. Any SCBA purchased after July 1, 1981, had to meet these requirements, and all SCBA used for structural firefighting had to meet these requirements by July 1, 1983.

Protective Clothing

Protective clothing that is used for structural firefighting and that was purchased after July 1, 1981, had to meet the requirements of 29 CFR 1910.156, and all protective clothing used for structural firefighting had to comply by July 1, 1983. Protective clothing includes head, face, and eye protection, hand protection, body protection, and foot and leg protection.

29 CFR 1910.1030

From a risk management perspective, nothing has impacted fire and emergency medical services (EMS) greater than 29 CFR 1910.1030, *Bloodborne Pathogens*. The passage of this standard has forever changed the patient-care methods for agencies providing emergency medical services. The standard requires that fire and emergency medical organizations perform a risk assessment of their operations from the standpoint that personnel safety is imperative. The risk assessment covers both emergency and nonemergency duties and operations. An exposure to a communicable disease can occur just as easily during the cleaning and decontamination of equipment at the fire station as it can during the delivery of patient care at an incident scene.

A common philosophy that must spread throughout the fire and EMS community is that all patients should be treated as if they have a communicable disease. Fire and EMS personnel must put their safety first in every situation. On every medical response, there is a potential exposure to communicable diseases such as

Laws, Codes, and Standards

hepatitis B, tuberculosis, and meningitis. Needle-stick injuries can lead to complications associated with human immunodeficiency virus (HIV) and progressing to acquired immune deficiency syndrome (AIDS). Knowledge and utilization of accepted precautions by all personnel are required, even though many see this as a break from a long-standing tradition in the fire service.

Risk management, as it relates to infection control, identifies such issues as personal protective clothing, use of mechanical resuscitation equipment, vaccinations for personnel, training and education, and the development of standard operating procedures.

This standard also affects other pre-hospital-care organizations. Examples are law enforcement personnel, housekeeping/custodial staff, mental health workers, and employees of other agencies who may be required to provide medical and/or first-aid treatment or cleaning and/or disinfection services.

FEDERAL MANDATES

The safety and health process in the fire service has greatly improved over the past 20 years and continues to do so. The tangible benefits of a proactive safety and health program are recognized through the reduction in the frequency and severity of accidents and injuries, lowered workers' compensation costs, improved protective clothing and protective equipment, and the development of sound and effective safety and health programs. Statutory laws apply to civil and criminal matters enacted by a body legally authorized to legislate requirements. A governing body can be a state legislative assembly, whose actions affect only one particular state, or it can be the United States Congress, which impacts all states and territories. After proposed legislation has passed the necessary hurdles and is passed by the legislative body and signed by the executive, the bill becomes law.

One example is the Superfund Amendments and Reauthorization Act (SARA) of 1986. Section 126 requires the Environmental Protection Agency (EPA) to issue an identical set of regulations, covering anyone not covered by 29 CFR 1910.120, *Hazardous Waste Operations and Emergency Response.* States that operate under federal OSHA would be required to comply with the requirements of the EPA, which are identical to OSHA's requirements. The importance of this to emergency response personnel is that both require the use of an incident management system, use of an incident safety officer at working incidents, and a health monitoring process for potential or real exposures of employees in a hazardous-materials incident.

The Ryan White Comprehensive AIDS Resource Emergency Act of 1990 has provisions that require emergency responders to be notified if an exposure to a communicable disease occurred during treatment of a patient. The notification process requires that the testing source or agency, such as a hospital, notify the affected employee directly. The Ryan White CARE Act protects the confidentiality of the affected employee. The final rule was released in 1994 by the Centers for Disease Control (CDC).

STATE LAWS

Each state or commonwealth has legislative assemblies that can pass laws that affect emergency services personnel. The process is similar to the action Congress takes in the development of a law. Once a bill is introduced to this lawmaking body, the bill travels through a lengthy process of approval by the assembly. It then travels to a chief executive, such as the governor, and, if approved and signed, it becomes law.

Two examples of laws affecting fire and EMS departments are the requirements that govern the emergency response of fire and EMS vehicles and a vehicle inspection program. Most jurisdictions have requirements for operating vehicles under emergency conditions, with defined responsibilities. This would affect emergency vehicles when passing through intersections that have traffic lights, stop signs, or yield signs or those vehicles encountering a school bus that is stopped and discharging passengers.

Development of an annual vehicle inspection program for all fire apparatus has become law in several states over the past several years. Due to an excessive number of firefighter fatalities caused by vehicle accidents resulting from poor vehicle maintenance or lack of vehicle maintenance programs, state legislatures have been prompted to mandate that all fire apparatus be inspected annually by a certified inspector.

An example relates to the Commonwealth of Virginia. Through a concerted effort by the career and volunteer fire service, the Virginia State Police, and the Virginia Department of Fire Programs, legislation was drafted to enact requirements for the annual inspection of all fire apparatus. An *ad hoc* committee formulated by the Department of Fire Programs went throughout the Commonwealth of Virginia educating fire service personnel on the importance of this legislation. Due to the overwhelming support by the fire service, the bill was passed by the state legislature and became effective July 1, 1992.

CONSENSUS STANDARDS

Standards are established by general consensus and become a procedure or document that can be followed or adopted. Nonregulatory organizations or associa-

Laws, Codes, and Standards

tions develop the standards from members who have requested to participate in their standards-making process.

These organizations or associations have guidelines that dictate the procedures for developing standards. The structure of the committees must be balanced so that one interest group cannot dominate the committee. Size or number of members is another consideration. The guidelines that govern the standards-making process relate to the process of incorporating comments or suggestions into committee standards, the method of developing a new document and the length of the process, and the revision process and the length of time taken to revise a standard, among other issues.

The poor safety record of the fire service over the years has been the leading reason for developing standards. These standards have addressed personal protective equipment and clothing, apparatus, hazardous materials, infection control, and other pertinent issues.

Consensus standards are not mandatory unless officially adopted by public authorities with lawmaking or rule-making abilities. Once a legislative body adopts and makes a consensus standard a law in whole or in part, the consensus standard becomes a mandatory requirement in that jurisdiction.

National Fire Protection Association (NFPA)

Since 1896, the National Fire Protection Association has been the world's leading nonprofit organization dedicated to protecting lives and property from the hazards of fire. NFPA is noted for its involvement in fire prevention and education programs and the standards-making process. The NFPA publishes over 286 nationally recognized codes and standards, and two of the best-known and most widely used standards are the National Electrical Code (NFPA 70) and the Life Safety Code (NFPA 101).

In 1986, at the NFPA Annual Meeting in Cincinnati, Ohio, the NFPA membership passed NFPA 1500, *Standard on Fire Department Occupational Safety and Health Program.* This standard has impacted the safety and health of the fire service greatly. The NFPA Technical Committee on Fire Service Occupational Safety and Health is responsible for the development of several standards that address specific safety and health issues. These standards are:

NFPA 1500	*Standard on Fire Department Occupational Safety and Health Program*
NFPA 1521	*Standard for Fire Department Safety Officer*
NFPA 1561	*Standard on Fire Department Incident Management System*
NFPA 1581	*Standard on Fire Department Infection Control Program*
NFPA 1582	*Standard on Medical Requirements for Fire Fighters*

NFPA 1500

The NFPA Fire Service Occupational Safety and Health Technical Committee's intent was to develop a user standard that would address various safety and health interests. NFPA 1500 serves as a voluntary consensus standard that details the needed requirements of a complete fire department safety program, and it meets or exceeds the criteria listed in the *Industrial Fire Brigade* standard of the OSHA requirements.

NFPA 1500 is comprised of 10 chapters plus 2 appendices. It was an original goal of this technical committee to eventually expand each of the chapter topics into its own document, and the list of companion documents above indicates that the process has begun. Incorporated into this standard are topics that deal with a wide range of safety and health interests. The chapters are as follows:

Chapter 1	Administration
Chapter 2	Organization
Chapter 3	Training and Education
Chapter 4	Vehicles and Equipment
Chapter 5	Protective Clothing and Protective Equipment
Chapter 6	Emergency Operations
Chapter 7	Facility Safety
Chapter 8	Medical and Physical
Chapter 9	Member Assistance Program
Chapter 10	Referenced Publications
Appendix A and B	

NFPA 1500 contains the minimum requirements for a fire department occupational safety and health program. It is intended to reduce the number and severity of accidents, injuries, and hazardous exposures involving personnel and equipment. It also recommends that all members (the term *member* being used throughout the standard to define a person involved with a fire department organization) abide by this standard whether they are career, part-paid, or volunteer. The technical committee's actions indicate that it believes that safety should apply to everyone who responds to emergencies, regardless of affiliation.

NFPA 1500 recommends that a fire service organization develop and maintain a written safety and health program. With this program, the organization must develop goals and objectives to prevent and eliminate accidents, occupational illnesses, and fatalities. A fire department health and safety officer must be appointed and would be responsible for the occupational safety and health program. The duties and responsibilities of the health and safety officer are defined in NFPA 1521, *Standard for Fire Department Safety Officer*. The development and functions of an occupational safety committee will be to conduct research, develop recommendations, and study and review matters pertaining to occupational safety and health within the fire department.

The original edition of the standard was released in August 1987, and the second edition was passed in May 1992, at the NFPA Annual Meeting in New Orleans.

The NFPA publishes standards that address protective clothing and equipment. These standards are the manufacturer's design criteria for protective clothing and equipment. As applicable clothing and equipment are ordered, the department should specify that the clothing and equipment meet these specifications. These standards include:

> Structural firefighting clothing and equipment, such as turnout coats, turnout pants, helmets, boots, and gloves;
> Protective equipment such as self-contained breathing apparatus (SCBA), personal alert safety systems (PASS), and rope;
> Protective clothing for personnel involved in wildland, hazardous materials, and crash, fire, rescue (CFR) operations;
> Protective clothing for personnel engaged in emergency medical operations.

Figure 4.2. A firefighter wearing full protective clothing and equipment for structural firefighting (photo by Martin Grube).

Figure 4.3. Fire apparatus must be designed with a fully enclosed cab and the number of belted seating positions specified (photo by Martin Grube).

NFPA has design criteria for various types of fire apparatus and equipment, which include:

> The manufacturer's specifications for pumpers, aerial ladders and platforms, and tankers;
> Design specifications for hose, ground ladders, and nozzles;
> The annual testing criteria for apparatus and equipment to ensure safety and operable conditions.

NFPA technical committees develop professional qualifications and competency standards for firefighters, fire officers, driver/operators, and fire instructors, and these standards are further discussed in Chapter 11. In addition, safety as it relates to live fire training, training centers, and training reporting procedures has been incorporated into standards. NFPA 1403, *Standard for Live Fire Training Evolutions,* was developed to address firefighter injuries and fatalities during live fire training exercises.

American Society of Testing and Materials (ASTM)

American Society for Testing and Materials is a private, nonprofit organization that develops standards for materials, systems, products, and services. It was founded in 1898 to provide these services for a variety of disciplines. Drawing from an immense pool of individuals, organizations, and disciplines, ASTM consensus standards focus on such topics as:

Medical devices
Security systems
Energy
Construction
Occupational safety and health
Protective equipment for sports
Transportation systems

Some of ASTM's efforts are directed toward consumer protection dealing with issues such as cigarette lighters, bathtubs and shower structures, children's furniture, and nonpowered guns.

Examples of the ASTM standards that impact emergency service organizations are:

ASTM Committee F-30 "Emergency Medical Services," which has generated the standards "Practice for Training the Emergency Medical Technician (Basic)" and "Guide for Structures and Responsibilities for Emergency Medical Services Systems Organizations"

ASTM D 3578, *Standard Specifications for Rubber Examination Gloves,* which includes requirements for sampling to ensure quality control, watertightness testing for detecting holes in gloves, physical dimension testing to ensure proper fit of the gloves, and physical testing to ensure that the gloves do not tear easily

American National Standards Institute (ANSI)

Serving as a national voluntary consensus standards-making organization in the United States, American National Standards Institute develops occupational and consumer standards. This organization is comprised of trade, technical, professional, labor, and consumer organizations, state and federal agencies, and individual companies that correlate standards development for these various groups when the need is determined by the Board of Standards Review.

Many of the standards or regulations developed through ANSI have been formulated due to the requirements set forth in the Occupational Safety and Health Act of 1970. Many of the safety and health standards are approved by standards committees under the direction of the ANSI Safety and Health Standards Board. Many of the approved consensus standards developed through the National Fire Protection Association become ANSI-approved standards.

Standard Operating Procedures (SOPs)

Standard operating procedures (SOPs) are written policies developed by an organization that specify specific methods or activities performed by members of that organization. These procedures affect only the operations of the organization that writes and adopts them. The requirements of these procedures should be based on recognized laws and regulations, and the organization should meet or exceed these requirements.

As with any procedure, if organizations expect all personnel to understand the requirements of that policy or procedure, it must be in writing. This written program will then provide the necessary components for the particular subject or procedure.

Influence and Effect of Laws, Codes, and Standards

A primary responsibility of an employer is to provide a safe and healthy work environment. The tangible benefits are the reduction of risk to the employees and the decrease in liability for the organization. The top administrator, such as the fire chief, has the ultimate responsibility for safety and health in the department. So that the process can be properly managed, this individual must appoint or designate a health and safety officer to be the program manager of the safety and health program.

The program manager must ensure compliance with all applicable laws, standards, and regulations developed and adopted by the organization. Complying with safety and health standards demonstrates a responsibility for safety and health of personnel. It sends a message illustrating to all members that the organization cares about them and their well-being. An organization that has effective operating procedures provides more-efficient and productive operations on a daily basis. Personnel are provided with the essential requirements to function in their assigned positions.

As an organization develops its policies and procedures, the laws, codes, and standards provide the necessary information to establish these policies and procedures. For example, NFPA 1500 provides the necessary components to establish a comprehensive safety and health program. Prior to the development of NFPA 1500, no other law, standard, or regulation addressed the safety and health needs for the fire service in such an inclusive manner.

Finally, the health and safety officer must develop a network that provides information on laws, standards, and regulations. As they are revised or developed, the health and safety officer can extrapolate this information to be included in the procedures for that particular organization.

Periodic Review and Revision Process

An organization must have a system in place to routinely evaluate, review, and revise procedures. Obviously, the health and safety officer and/or the occupational safety and health committee will be responsible for safety- and health-related policies. Also, safety and health procedures should be audited by an external resource at least every 3 years, as outlined in Chapter 10. This audit should be conducted by a qualified individual from outside the fire organization, because outside evaluators provide a different perspective, which can be constructive.

The job functions and duties of the health and safety officer or the occupational safety and health committee must be routinely reviewed and revised based on new assignments or responsibilities. This also determines whether the health and safety officer is concentrating on the primary functions of the position. If a health and safety officer position is established, the occupational safety and health committee must still function as part of the organization's safety and health process.

Conclusion

Laws, codes, and standards must be an integral part of the organization's operating procedures, both emergency and nonemergency. The primary intent of these regulations is to protect and enhance the safety and health of personnel. However, this can be taken a step further to improve the services provided to customers. Compliance with laws, codes, and standards will lead to better procedures, enhance the public's perception of the organization by providing a more efficient and effective operation, and help to ensure the safety of the members of the organization.

PART 2 | *Comprehensive Risk Management Plan*

PROFILE

Chapter 5: The Management of Risk

MAJOR GOAL:

To be able to establish risk management goals and objectives, and to utilize a plan for achieving them

KEY POINTS:

- Understand that from a risk management perspective, there are four categories of losses:
  - Personnel
  - Property
  - Legal liability
  - Time element

- Clearly define the goals and objectives of the risk management effort.

- The classic risk management model has five steps, as listed below. Understand the relationship between them before initiating the first step.
  - Risk identification
  - Risk evaluation
  - Establishing priorities
  - Risk control
  - Monitoring of program

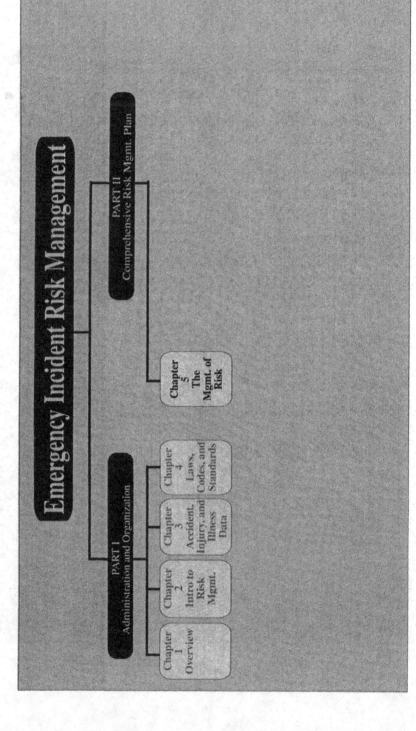

Chapter 5

The Management of Risk

THE PROCESS OF MANAGING RISK

Before embarking on a lengthy discussion of managing risks, we need to get one point straight: responding to emergencies is a dangerous business. Those who respond to emergencies face an inordinate number of risks daily, and it is essential that those risks be controlled as effectively as possible. Risk management is not a magic bullet that will cause all risks and dangers to disappear. When used appropriately, it can certainly help to reduce the impact risks might have on an organization.

Risk is defined in many ways. Technically, as was stated in Chapter 2, it is a measure of the probability and severity of adverse effects. However, that definition does us little good in trying to manage risks, so, for our purposes, we shall consider risk to be any chance of a "loss" occurring. Essentially, it is the probability of anything that poses a threat to the organization occurring. Risks that remain uncontrolled will result in a loss, and these losses will be a drain on the organization's resources. Losses fall into the following four broad categories:

Personnel loss Just as the name implies, this is the loss of the services of the people in the organization, through death, illness, disability, etc. This also includes any costs associated with replacing workers who are lost.

Property loss This refers to damage to or complete loss of property. (For example, the burning down of a fire station would be considered a property loss.) Damage to property is included in this category because the use of the property would be lost until the property is repaired or replaced.

Legal liability Liability is any legally enforceable obligation. Losses in this category are almost boundless, ranging from the results of the use of allegedly flawed firefighting strategy (reasoning such as: "the building wouldn't have been a complete loss had the fire department performed better"), to injuries arising out of a vehicle accident deemed to be the fault of the department's driver.

Time element This characterizes a loss whose ultimate value is determined by an element of time. A good example of a time element loss is the loss of rental revenue a landlord would suffer if, for some reason, the tenants were unable to inhabit a building he owned.

For almost all losses, financial resources are the ones most easily measured and, therefore, most often quoted in determining the seriousness of a loss. With the heightened awareness in recent years of the liability "crisis" facing organizations, risk management has started to gain some prominence. After all, a dollar not spent on a liability loss can be used elsewhere.

However, let's move away from consideration of financial resources and turn our attention to personnel. When people are injured, become ill, or die, there are far greater consequences than loss of money. The historical concentration on liability issues may have come at the expense of people. Risk management can have as dramatic an impact on the health, safety, and well-being of an organization's people as it can on the bottom line.

We defined a risk as anything that poses a threat to the organization, its people, or its resources. It follows then that risk management is the effort made to control those threats. There are several techniques for controlling or managing risk, and this part of the book will examine them in order of priority, from most effective to least effective.

CHOICES

Risk management is about choices—decisions that have to be made, often on a daily basis, but that depend upon the review and analysis of quality information, as discussed in Chapter 3. Many of the decisions are not easy, and data that have been converted to information will not always provide a clear-cut choice.

For example, suppose that a decision has to be made concerning the purchase of a smoke ejector. A larger ejector can move more smoke, but it takes up more room in the compartment, and requires two people to handle it. On the other hand, a smaller ejector will be less effective at an incident, but can be stored and handled more easily. The information is clear, but the decision is not.

When the decision is made, the results of that decision will impact the members of the organization in some form or another. It is up to the individual who must make the decision to understand and balance all the factors and their consequences, and then act. Is this a risk management decision? Not directly. Does it impact the health and safety of members? Absolutely.

Regardless of the techniques chosen or the methods used to apply them, risk management is a process, not a static event. As with any process, it is dynamic, ever changing, and in frequent need of review and revision to ensure that it remains current and effective. We stated above that risk management is not a magic bullet; the process cannot and should not be ignored once it is implemented.

Goals and Objectives

At the outset, it is important to outline risk management goals and objectives. Goals establish the purpose(s) for a process, and the objectives associated with each of the goals provide a road map for achieving it. These goals and objectives should not be unlike others established within the organization. They should be reasonably attainable, possibly requiring a little "stretch" to be accomplished. They should be clearly stated and understood, and be measurable. In other words, they should make clear what is to be achieved, why it is to be achieved, and how progress in achieving it is to be measured.

Although such an exercise may appear to be straightforward, it is important in helping to ensure that the risk management process has a distinct, clearly understood, and stated purpose, and that any actions taken keep it focused on that purpose.

There are several risk management goals that are nearly universally recognized. They include the following, which are not listed here in any order of priority:

> Survival of the organization (given that a severe loss could potentially put an organization out of busines)
> Lower costs
> Limited interruptions to operations
> More stable or predictable outcomes from operations
> Continued growth
> Positive public relations
> Compliance

Although general in nature, these goals can be applied to any organization, regardless of the type of business. Their genesis is the positive outcome that may be realized by implementing effective risk management strategies.

Some sample risk management goals applicable for a fire department or other emergency response organization are listed below. The objectives established for achieving the goals will be the result of the application of the classic risk management model, which is described in the following chapters. The sample goals are to

Reduce the number of worker injuries
Control the costs associated with apparatus accidents
Establish a more accurate, predictable budget for insurance costs
Improve overall firefighter safety and health
Demostrate to customers a commitment to safety

RISK RETENTION

Despite our best efforts to the contrary, some risks will have to be retained or assumed. Any losses associated with those risks are consciously accepted by the organization, rather than transferred or avoided, as we shall discuss later in this section of the book. In real terms, this means that losses will be paid directly by the organization, either out of operating funds or through some other funding method.

It is the risks that we ultimately retain that require control measures. We shall discuss a variety of controls, but none can be 100% effective, or make the risk completely disappear. When all is said and done, the organization will find itself dealing with an element of risk, and that is part of doing business. Risks that are retained are not necessarily uncontrolled; because they are known to the organization, strategies can be implemented to limit any losses associated with them.

ADMINISTRATIVE RISK MANAGEMENT VS. EMERGENCY INCIDENT RISK MANAGEMENT

Any group that responds to emergencies, whether it be a municipal fire department, private fire department, industrial fire brigade, or emergency medical service provider, faces some unique risk management challenges. Unlike most other businesses (and the above-mentioned groups *are* businesses), they must deal with two distinct risk-producing situations: routine day-to-day operations and emergency incidents.

Businesses and individuals alike must manage nonemergency risks. There is nothing that distinguishes an organization that responds to emergencies from one that doesn't when no emergency is in progress. Firefighters may fall down stairs at the station just as office workers may; emergency medical technicians may suffer back injuries while moving supplies just as warehouse workers may. Therefore, there is a great deal of similarity and consistency in the techniques for managing these risks, regardless of the organization.

When an emergency call comes in, the circumstances change. No customer calls for emergency assistance because something is going *right*. Customers plan to control the risks associated with their emergencies by inviting someone else onto the premises to handle them. For these people, the most effective way to manage emergency incident risks is to dial 9-1-1. The members of an agency that responds have no such "out." The buck stops with them, and they must have the tools available to handle the incident effectively and safely.

Another, more dramatic factor that distinguishes between nonemergency and emergency incident risk management is *time*. Whatever we choose to call them—routine risks, administrative risks, or nonemergency risks—there is typically no immediate constraint on the time we use to decide what to do about them. The decision on how much fire insurance to purchase for a fire station can be made over time, with the deadline usually determined by the insurance company, based on the expiration of the existing policy.

At an emergency incident, one does not have the luxury of time. Numerous risks exist for the members who have responded, for their apparatus and equipment, and for any civilians who may be involved in the incident. To be truly effective, the decision as to which risks are of highest priority and what to do about them based on the resources available must be made quickly, if not instantly. The stakes are high. The decisions determine whether lives are saved or lost, and there is no second chance to be right.

CLASSIC RISK MANAGEMENT MODEL

In the next several chapters, we shall outline the classic risk management model. That model presents a systematic approach for identifying risks and planning for their control. The methodical process for making decisions can be utilized not only for the nonemergency risks that all organizations must address, but also for the risks associated with the response to and mitigation of an emergency incident. The factors at each incident will always vary, but, as has been known for years, continual training in all aspects of the approach will yield the best, most consistent results possible.

The model has five basic steps, as listed below. Each one depends upon information generated at the preceding step, so it is important to evaluate each one before moving on to the next. The following chapters will discuss these steps in depth, and, when taken as a whole, will provide the framework on the basis of which individual organizations can develop their own risk management plans.

Step 1 *Risk identification:* This states what might go wrong, and involves various means for identifying the risks the organization faces or might face, and where to turn for assistance.

Step 2 *Risk evaluation:* This determines, comparatively, which risks present the greatest concern, and why.

Step 3 *Establishment of priorities:* This determines the difference between identified risks that will require immediate action, and those that can wait.

Step 4 *Risk control:* This, which is arguably the most important step, states what can be done to manage, control, or eliminate the risks that have been identified and evaluated.

Step 5 *Monitoring:* This builds into the process a means for evaluating the effectiveness of the control measures implemented. Modifications can then be made as necessary.

Conclusion

In this chapter we have addressed risk and the consequences associated with a risk going unmanaged. When particular risks are not controlled, four types of losses are possible: personnel loss, property loss, legal liability, and time element.

In order to begin to manage risks it is essential to establish reasonable goals for a risk management program and to outline objectives that will help the organization to achieve the goals. This has been described as identifying where one is going prior to setting out. Otherwise, it will never be known whether the destination achieved was the correct one.

Once the goals and objectives have been identified, the classic risk management model can be applied as a method for achieving them. The model is outlined and analyzed in the next several chapters, and it will be a more useful tool when used with the background information presented in this chapter.

PROFILE

Chapter 6: Risk Identification

MAJOR GOAL:

To be able to effectively identify the risks to the organization, and utilize a method to document them

KEY POINTS:

- Understand why risk identification is the most important step in the risk management process.

- Understand that the risk identification process needs to be structured. For example, risks can be grouped by:
    - affected group
    - organizational operation / division
    - line of insurance coverage

- Document all identifiable risks.

- Understand that there are numerous sources of information available for use in this step. Use any and all that apply.
    - past-loss data
    - members of the organization
    - trade / membership organizations
    - consultants / outside experts
    - on-line services
    - trade publications

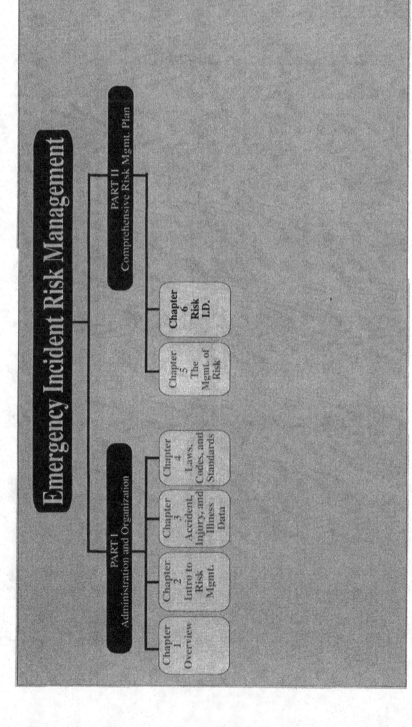

Chapter 6

Risk Identification

INTRODUCTION

The subtitle of this chapter could be "What Might Go Wrong?" Risk identification is the first step in the risk management process. If we truly consider risk management to be a process, and liken it to assembling building blocks, risk identification would be the bottom blocks. Everything that follows is supported by the results of this step.

Once the risks to the organization have been identified, an action plan for controlling or eliminating them can be developed and instituted. That also makes this a crucial step, because one cannot develop control plans for risks that have not been identified. The steps in the process are truly interrelated.

Before risks can be identified, it is important to review the definition of what a risk is. In Chapter 5, we decided to consider a risk to be any chance of a loss occurring, with losses falling into the four broad categories of personnel loss, property loss, legal liability, and time element. A loss can range from a minor dent in a vehicle to an employee fatality. If there is a chance that any of these can occur, then there is a risk. If there is truly *no chance* of a loss occurring, there is no risk.

In identifying risks, it is not so important to consider the type or extent of the loss, because those factors will be considered later. For example, if the risk identified is vehicle collisions, then damage to the vehicle would be a property loss. Because the vehicle would be out of service during repairs, there might also be a time-element loss. This loss would include a financial component (cost of repairs, increased insurance premiums) and reduction in productivity (services temporarily unavailable for delivery). For now, however, simply identifying the risk of collisions is sufficient.

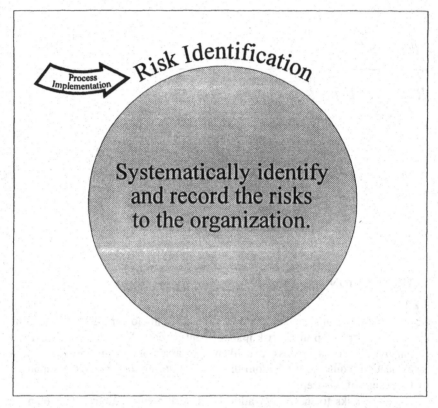

Figure 6.2 Risk Identification is the first of several interrelated steps in a continual process.

Just as there are varying types of losses, there are different types of risks. Many of the definitions are more technical than is required for our purposes, so in our risk identification process, we shall consider only *actual* and *potential* losses.

Actual losses are ones that have occurred in the past and can be reasonably expected to happen again. Potential losses are ones that have not yet happened but are possible. Although a serious fire in a fire station has not yet occurred, the potential exists, and this risk must be identified during the identification process.

There are a variety of methods that can be utilized to systematically carry out the risk identification process. We shall explore several, but, ultimately, it is up to each individual to find a method that works best for him or her.

The key word is *systematic*. A disorganized, illogical effort will likely provide an incomplete product, and also be a real source of frustration for those involved. Initially, it is essential to identify all reasonable risks, because comprehensive control efforts will be orchestrated based on the results of this step.

As we shall discuss later, periodic monitoring and updates of the risk management program are necessary. These usually occur after the program has had some time to mature. On the other hand, if it becomes necessary to continually revisit any one of these steps during the initial stages of implementing a program, that may be an indication that the work on that step is incomplete. Because the risk identification step forms the basis for all of our later work, we can avoid that frustration by being thorough and systematic initially.

There is no absolutely right or wrong way to approach this exercise. The methods outlined provide guidance on some means of organizing the effort, using some common denominator. It is likely that there are other systems that would be as effective, and entities are encouraged to use their initiative to make this effort as productive as possible.

As examples, we shall explore methods that list risks by affected group, by organizational operation/division, and by line of insurance coverage. These are typical for almost all organizations, and serve as a valid reference point. In reviewing the various methods, observe how a single identified risk may be listed differently in each one.

A good place to start, and one that may help with the decision on what method to use, is past-loss history. What has historically gone wrong? The data may well fall into some natural categories or allow the identification of trends. By building on this as a start, it may be easier to fill in the blanks.

It also holds that any losses that have been occurring consistently will continue to happen until intervention occurs in the form of control measures. If a particular organization has been experiencing an average of 18 back injuries a month among its workers for the last four years, it is likely that there will continue to be 18 back injuries each month until something is done to alter the trend.

RISK IDENTIFICATION METHODS

Affected Group

This approach lists risks to the entity in terms of who or what will be affected. In most organizations, these groups typically would be: personnel, facility, apparatus, and equipment.

The risks faced by *personnel* are many and varied. For firefighters, there are the well-known ones, such as exposure to fire and fire products. There are also risks associated with exposure to hazardous materials, stress and related disorders, back injuries, and the risks presented by special operations such as confined-space entry and rescue. A review of past injuries and illnesses will indicate where the risks unique to a particular entity have occurred and will likely continue to occur to personnel.

Risks to the *facility* include fire, flood, hurricanes, and other natural causes. Damage due to acts of vandalism, civil unrest, and terrorism would also be included here.

When department administrators think about risks, *apparatus* accidents are frequently the first that come to mind. In many entities, these may occur more often than any other type of loss. Vehicle collisions would top the list, followed by damage caused by acts of vandalism, theft of the apparatus, and other damages, such as broken windows and blistered paint.

The last group is *equipment*. Items that the department uses are subject to loss, damage, theft, and breakdown that results in loss of use.

Organizational Operation/Division

Many administrators find it easier to assign or group losses, based on the operation or division those losses are most likely to affect. A large municipal fire department, for instance, may have suppression, prevention, communications, maintenance, and public information divisions, among others. Many of the risks associated with operating each of those divisions may be unique, so a system that groups losses the same way may be very effective and reasonable. In a corporation, the same logic can be applied, using the company's operating divisions. At a single site there may be production, shipping, administration, and warehousing.

This method of assigning or grouping losses can be effective, but is dependent upon the needs and structure of the entity being evaluated. For seamless organizations with indistinct departmental boundaries, or ones that operate using teams rather than assigned work groups, this method may not be as easy to utilize.

Line of Insurance Coverage

We shall discuss some of the basic insurance coverages applicable to organizations in Chapter 9. In this phase of the process, the lines of coverage usually utilized for identifying risks are listed below. Entities will have other lines of coverage, and each protects against specific risks, but the primary ones we will address are those involving workers' compensation, general liability, auto (both liability and property damage), and property.

SOURCES OF INFORMATION

Regardless of the method used for identifying risks, there are several readily available sources of information to use. Appendix A lists some of the more common risks and suggested control measures. Other sources that we shall discuss include the following:

Risk Identification 75

    Past-loss data
    Members of the organization
    Trade/membership organizations
    Consultants/outside experts
    On-line services
    Trade publications

Each source can provide information which will allow risks to be identified, and ultimately, evaluated. How that data is "classified" depends on the identification method chosen.

Past-Loss Data

As we mentioned earlier, anything that has consistently gone wrong will continue to go wrong until steps are taken to mitigate the situation. Therefore, trends in an entity's loss experience will yield valuable information in the risk identification process. Chapter 3 includes an in-depth discussion of data, their availability, and how they can best be utilized.

Data specific to the organization should be easily available from two primary sources. The first is records of losses that are maintained locally. Knowledge about the organization is crucial for effective conduct of business, and past-loss experience is one of those pieces of information that the administrator needs to have available. If data about previous losses are not available locally, some method for collection and eventual retrieval of that information needs to be instituted.

Assuming that the data are available, this is the first place to start. Information may be stored in different ways or locations, but someone within the organization probably has access to it. Examples of internal information are apparatus accident reports, summary personnel accident and injury data, and maintenance requests for equipment.

Summary personnel accident and injury data must be stressed. Much of this information may be confidential, and there are stringent requirements regarding how and where it may be maintained. For risk identification purposes, descriptions of types of accidents and injuries and how they occurred are adequate. It is not important to know the names of affected individuals or details on the type of treatment received.

A second source for the information is the insurance carrier or carriers for the organization. If different insurers handle different lines of coverage for the organization, then each will have to be contacted. However, it is part of an insurance company's service to provide loss data.

Frequently, an insurer who receives a request for past-loss experience may believe that the client is shopping for the same insurance from a competitor. If that is the case, the insurer will be in no rush to provide the information, so it may be wise to advise the company of the reason for the request. This may also prompt the insurer to arrange to have someone from the organization hand-deliver the data and assist with evaluating them.

Learning which insurance company handles which line of coverage may be easier than it sounds. It is likely that within each organization there is someone who deals with insurance matters. If so, this individual may be able to provide the names and addresses of the insurers so that the request can be made.

If the organization uses an insurance broker or agent to place the coverages, this individual can make the requests on behalf of the organization. A broker is usually a local business person who has access to several insurance carriers. He or she can provide some guidance on insurance-related matters, and this type of request for information is definitely within the scope of a broker's services. In this instance, the use of a local broker will make it easier to follow up if all information is not received in a timely fashion, because only one, local call will be required.

Members of the Organization

To list potential losses, which are the ones that have not happened but that may, is a difficult exercise. Reviewing past-loss information to identify future incidents is relatively easy, but predicting what hasn't happened is not!

The members of the organization closest to the action are frequently in an excellent position to provide valuable information for this part of the exercise. They have a clear understanding of the day-to-day operations, and can usually predict, with a high degree of accuracy, accidents that are waiting to happen.

The format for harnessing this information will depend largely on the dynamics of the particular organization. Safety committees may be a viable forum for collecting the data, as may employee organizations. The purpose for gathering the data is to prevent accidents, injuries, and other incidents, and that purpose should be clearly communicated to all involved parties, to prevent any misunderstandings. This is one area in which the affected groups can and should come to terms, because the results of the efforts will benefit all.

Because a discussion about potential losses may not be an easy one to get started, some sample questions to stimulate thinking are listed below.

1. What close calls has anyone had recently?
2. What do you believe is the most common cause of injuries in our department (company, division, etc.)?
3. What is the biggest threat to the health and safety of the members of our department (company, division, etc.)?
4. If you could do just one thing to improve the safety of our department (company, division, etc.), what would it be?

Risk Identification

5. What would be the most effective way of preventing future apparatus collisions?
6. If you were personally responsible for ensuring that our portable equipment remained in top condition, what are the first three things that you would do?

Trade/Membership Organizations

The cost of insurance, especially health and workers' compensation due to the medical component, is a critical issue facing organizations. So, too, are personnel health and safety concerns. As a result, trade organizations are working hard to provide services to their members to assist them. Information about loss experience is usually available regarding individual members of the association or in aggregate form. Either way, this is an excellent way of determining what has happened to other, similar entities.

Appendix C includes names and addresses of several national trade associations that may provide valuable information to members. Some of these include the Public Risk Management Association (PRIMA), the Risk and Insurance Management Society (RIMS), the International Association of Fire Chiefs (IAFC), the International Association of Fire Fighters (IAFF), the National Fire Protection Association (NFPA), and the National Safety Council (NSC).

In addition to these national organizations, there are many local and regional ones that may also be of assistance. For instance, most states have associations of fire chiefs whose purpose it is to address the collective needs of members. If all chiefs are struggling with this exercise, the association may well be able to provide assistance to prevent each chief from having to "reinvent the wheel," so to speak.

Membership in such organizations will provide access not only to the formal services offered, but also to other members. The networking opportunities presented by membership and subsequent meeting attendance are myriad. Just discussing a problem with a peer may shed light on a solution not yet considered. At this stage of trying to identify risks, this kind of communication is invaluable.

The National Safety Council annually publishes *Accident Facts*, which highlights accident data in a variety of formats. This publication is also available from local safety council organizations. In the United States, there are currently 88 accredited chapters of the National Safety Council, and there is one in Quebec. These local connections may be able to provide a great deal of information.

Both the International Association of Fire Fighters and the National Fire Protection Association publish annual summaries of firefighter deaths and injuries. There is much that can be learned from studying the circumstances of the mishaps of others, so this information should not be overlooked.

Consultants/Outside Experts

There are professionals available to assist with this segment, as well as others, of the risk management process. Frequently, the personnel who are attempting to identify the risks to the organization must deal with challenging questions or situations, and the expertise and judgment of an outsider may be appropriate.

Many consultants will charge a fee for their services, but others are available free of charge, typically through an insurance company. Understanding the resources available, and how to access them, will again prove to be a cost-effective exercise if the services of a consultant become necessary.

On-Line Services

Many organizations have at their disposal today access to global information. On-line services—so called because the user is electronically connected "on-line" with thousands of others—have exploded on the scene over the last decade. Many of these provide a focus for needs and can be accessed for information.

As with any burgeoning field, this list can change from month to month, and even from week to week. Those who intend to utilize on-line resources are well advised to stay current on the various services and their offerings, in order to get the most out of their efforts. In addition, many of these are offered on a fee-for-service basis, so a little research prior to connecting will help to ensure a cost-effective exercise.

For example, "ICHIEFS," which is offered as a service by the International Association of Fire Chiefs, allows subscribers to communicate with each other. A simple query on ICHIEFS about loss experience is sure to yield a wealth of data.

"RIMSNET" is offered through the Risk and Insurance Management Society. Primarily concerned with more technical aspects of risk management, this too should provide a great deal of relevant information.

Trade Publications

Just as on-line services link subscribers who have common interests, trade journals have target audiences. Some of the more common trade journals in the emergency services arena include the *Journal of Emergency Medical Services (JEMS)*, *Fire Engineering*, and *NFPA Journal*.

From a more purely risk-management standpoint, there is *Risk Management* (published by RIMS), *Professional Safety* (offered by the American Society of Safety Engineers), and *Public Risk* (offered by PRIMA). Any of the aforementioned periodicals routinely carries articles about insurance, accidents and injuries,

SAMPLE RISK MANAGEMENT PLAN

IDENTIFICATION	FREQUENCY	SEVERITY	PRIORITY	ACTION REQUIRED / ONGOING	CONTROL MEASURES
A. Sprains and strains					
B. Cuts and bruises					
C. Stress					
D. Terrorism and the workplace					

Figure 6.3
A suggested format for recording decisions made during the risk management process. Four risks have been identified.

and safety. A trip to a good-sized reference library will often prove to be worthwhile in the quest for good, relevant data.

Although no trade journal can claim to be immediately interactive, as an on-line service can, there are opportunities to communicate with other readers. The editor frequently presents controversial viewpoints to stimulate readers, and return letters to the editor are published for others to see. These journals also welcome contributions of material by readers, which provides another opportunity not only to learn, but also to share information that may assist others.

RECORDING OF FINDINGS

We are still in the first step of a multi-step process. In identifying risks, simply making a list of them is appropriate. Assuming they are organized in some fashion, it will be easier as we progress if the list is also organized.

The Sample Fire Department Risk Management Plan shown in Figure 6.3 offers a suggested format for starting such a list, along with sample entries. In this figure, the Identification column is highlighted. As we progress through the model and more information is generated, we shall insert the appropriate entries in the remaining columns.

CONCLUSION

Risk identification is the first in a multistep process for managing risks. Many argue that it is the most important step because of the difficulty, if not impossibility, of managing risks that have not been identified.

The risk identification process will be more productive if it is conducted in an organized fashion. Risks can be grouped in several different ways, and each organization should decide which method will be most effective for their situation. The purpose for using some sort of system for performing the risk identification is to ensure that no risks are overlooked, because such oversight would make the next several steps in the process that much more difficult.

As the risks are identified, they should be recorded. At that point, they do not have to be listed in a particular order of importance. That order will be determined when the risks are evaluated, which is a topic that will be discussed in the next chapter.

PROFILE

Chapter 7: Risk Evaluation

MAJOR GOAL:

To understand how to evaluate the risks that have been identified, and record the results

KEY POINTS:

- Understand that frequency is how often an accident, injury, or other loss occurs.

- Understand that there is no universally established frequency scale. The frequency is what it is; the determination of whether that frequency is acceptable is based upon judgment and experience applied to local circumstances.

- Understand that severity is a determination of how serious the impact of a loss will be, and is measured in several different ways.

- Understand that severity, like frequency, has no universal scale. Judgment and experience again will be required to determine the potential seriousness of losses associated with a risk.

- Understand that neither frequency nor severity alone provides a valid measure for establishing priorities for action. The two factors need to be evaluated together.

- Continue to build the risk management plan that started with the risk identification process by incorporating the evaluation for each risk.

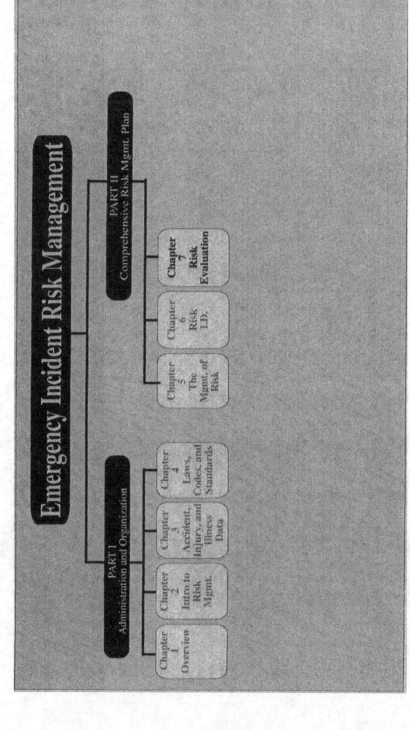

Chapter 7

Risk Evaluation

INTRODUCTION

In Chapter 6, we explored several systematic methods for identifying the risks that threaten the organization. During this identification process, we paid little heed to the potential seriousness of each risk. In this chapter, we shall examine some factors that can be used to evaluate the risks and thus allow us to begin to set some priorities for action. Determining the seriousness of a risk to the organization can be a challenge. After all, there is no one universally accepted criterion for "seriousness." If the question is whether it is hot or cold outside, we simply look at a thermometer to decide. If the question is whether a risk is serious or not, there is no such device we can check!

EVALUATION MEASURES

For each of the risks identified, we can apply two measures to try to establish their impact on the organization. The first measure is *frequency,* and the second is *severity.*

Frequency is an evaluation of how often a particular incident may occur or is likely to occur. Severity is an evaluation of the consequences once an incident does occur.

Although these factors may seem easy to evaluate on the surface, even at this point judgment becomes important. Using the thermometer example again, is a

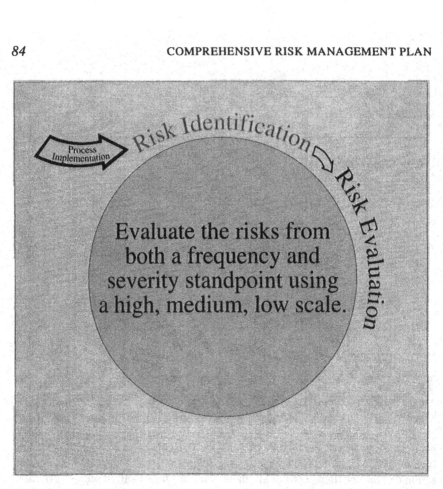

Figure 7.2. Risk Evaluation is the second step.

temperature of 55 degrees Fahrenheit cold or hot? If it is midnight at the North Pole, 55 would probably be considered a heat wave. On the other hand, in the Sahara Desert at noon, 55 may well be sweater weather! The point is that although the number may be the same, local circumstances can have a dramatic impact on how the number is interpreted.

Frequency

How often will an accident, injury, or other loss occur as a result of any of the risks identified in the first step? This is information that is available from the research conducted while the risks are identified. Historical data from the organization's own losses will help determine frequency.

As described in Chapter 6, an organization that typically records 18 back injuries a month will probably continue to do so until steps are taken to alter that trend. The

Risk Evaluation

question that remains, however, is whether 18 back injuries per month present a serious frequency problem, or whether 18 is actually a reasonable number for that organization. We shall examine that question in greater detail in Chapter 8, "Establishing Priorities." For now, however, let's assume that the number of back injuries is a fact, with no positive or negative connotation connected with the number.

As mentioned above, the frequency evaluation also applies to potential incidents. The reasonable question to ask is: How can we determine the frequency impact of something that hasn't happened, but that might? The answer will become clearer as we work further into the process and evaluate both frequency and severity factors simultaneously. The fact that a fire station has not burned down does not mean that it won't; however, such an event is a risk that must be identified even if the frequency of its occurrence will, it is hoped, be extremely low.

The next column in the Sample Fire Department Risk Management Plan that we initiated in Chapter 6 is titled "Frequency." The entries for this column will consist of the evaluation of the frequency of each of the risks already identified and recorded. A measure such as low (L), medium (M), or high (H) is appropriate for this. And this is where professional judgment and tolerance for losses by the entity's administration and personnel comes into play.

A total of 18 back injuries in a municipal department that has 1,000 members is probably reasonable, but that same total in a very small department would be considered dreadful. At the same time, the administration and personnel in the 1,000-member department may decide that the only tolerable number is zero injuries, so the 18 back injuries may be assigned an "H" for them as well.

The assignment of a frequency determination does not need to be scientifically justified. It is based on the judgment and experience of the people making the determination, and may be modified as additional information is gathered. However, it will serve as a valid factor as we begin to establish priorities for action.

Severity

The severity of the occurrence is the second half of the evaluation process. What constitutes a severe loss? There are several factors that can be used, and we shall highlight several of them. It becomes a moral and ethical question as to what the appropriate priority for each factor should be. The cost of an occurrence is often used as a determinant for severity, but is the cost of a heart attack, for example, really the most important aspect of it? Many would also ask about the impact on the individual (how is he or she doing?), the likelihood of continuing in the career, the loss of the individual's valuable experience to the organization, and so on. Therefore, we shall list the factors, with no opinion on which is most or least important.

Cost

How much does it cost the organization if a given incident occurs? Cost is measured on several different levels, and we should try to anticipate all of these costs. First, there are direct costs such as medical bills, overtime to fill vacant shifts, cost of repair/replacement, and insurance deductibles for apparatus accidents. These represent real money and costs that are relatively easy to determine.

The indirect costs may be more difficult to quantify. Several examples of indirect costs are listed below. The list is not all-inclusive, but it provides a realistic view of the factors that can be considered.

Lost productivity while someone is out due to an injury;
Additional costs associated with training or retraining a replacement worker;
Lower efficiency if reserve apparatus is pressed into service;
Higher insurance premiums;
Lower productivity due to use of a less experienced replacement worker;
Increased paperwork associated with the incident;
Lost productivity while remaining workers cope with the circumstances surrounding the occurrence.

An analogy for estimating indirect costs is that of a resting alligator, as shown in Figure 7.3. The part of the alligator that is visible represents direct

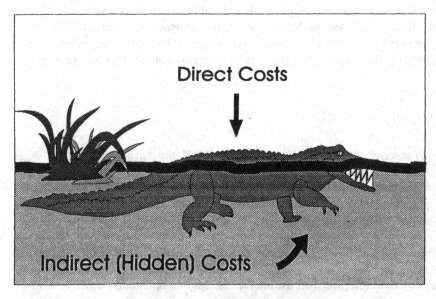

Figure 7.3. Just like an alligator, the most dangerous part is hidden.

Risk Evaluation

costs. The larger part of the alligator, that part that has the big teeth but that is not visible because it rests below the waterline, represents the indirect costs. Studies have concluded that indirect costs are as much as six to ten times greater than direct costs. Therefore, a sprained ankle with direct costs of $5,000 (medical, indemnity, overtime, etc.) may have indirect costs of $30,000 to $50,000. Needless to say, money can be a strong determinant of the severity of an incident.

Impact on the Organization

Impact can be measured in several ways. If a fire station burns down, the event would not only be expensive, but would also have a dramatic impact on the members of the organization, from an emotional standpoint. For many, the station may be a home away from home, and losing it would take a toll on them.

The nature of the loss may also determine how an organization can be impacted by it. If a piece of equipment or other resource critical to the operation is lost, damaged, or destroyed, the ability to deliver service will be compromised. An example might involve a key component of the dispatch console that affects the notification of personnel to respond to an emergency. If the console is disabled, some backup means of notifying personnel will have to be utilized until repairs can be made. Although the cost of the repairs may be minimal, the loss may be considered high on the severity scale, due to the adverse impact.

Time/Resources Required for Rectification

A risk may be considered to happen infrequently and be inexpensive to rectify, but may still be considered severe if an inordinate amount of time or substantial resources are involved in making it right. Many will point out that time is money, and that these considerations should fall into the cost category. However, this factor for evaluating severity may combine both cost and impact on the entity.

For example, consider a rural organization whose sole transmitter is located on a remote mountaintop. Should the transmitter malfunction, several resources must be utilized to make the repairs. First, a qualified technician must gain access to the transmitter and, once there, diagnose the problem. Once the problem has been diagnosed, appropriate parts will have to be acquired, and then a return trip to the transmitter must be made to rectify the problem. While all this is going on over a period of hours or even days, the agency may have only limited communication capabilities. This could dramatically affect the service to customers and place employees in danger.

In this example, none of the parts needed for repair were expensive, but because of the time and resources required to rectify the situation, this risk could be rated infrequent yet severe.

After all the factors are available for consideration to determine severity, we now condense everything to make an evaluation of severity as low (L), medium (M), or high (H). In the Sample Fire Department Risk Management Plan, we have now entered the evaluations of both frequency and severity, and the appropriate columns are highlighted, as shown in Figure 7.4.

FREQUENCY AND SEVERITY CONSIDERED TOGETHER

To draw reasonable real-world conclusions, the frequency and severity determinations that have been made must be considered together. In so doing, the significance of each of the risks that have been identified can be more accurately assessed, and this will be shown to be important when we attempt to establish priorities in Chapter 8. Although we address frequency and severity individually for discussion purposes, they are typically evaluated simultaneously.

Figure 7.5 presents a graphic method that can be used to illustrate and then conduct this evaluation. Frequency will be charted on the vertical axis, or y-axis, with severity on the horizontal axis, or x-axis. Both start at 0 and increase from there.

Note that "near misses" are listed along the y-axis. These are incidents that occur but that do not result in a loss. When a worker falls but is not injured, a near miss has occurred. These incidents are important to note, and possibly to record, because they may indicate a weakness in the risk management program or some condition that needs to be addressed in order to prevent a recurrence. There may be a very high frequency of these, but the severity will be zero because there really is no loss to measure.

As the Sample Risk Management Plan is completed with the list of identified risks, and the determination of frequency and severity factors is made, this information can be entered on the frequency/severity chart. Using the information from the sample plan in Figure 7.4, both the frequency and severity of each risk (designated A through D) have been entered on the chart to create the resulting scatter diagram shown in Figure 7.6.

For those who will complete the charting exercise, it is important to note two things:

1. It is impossible to assign definitive values to the axes. The chart will be different for every body of data, so a sliding scale determined by local circumstances must be applied to make the diagram exercise effective.

SAMPLE RISK MANAGEMENT PLAN

IDENTIFICATION	FREQUENCY	SEVERITY	PRIORITY	ACTION REQUIRED / ONGOING	CONTROL MEASURES
A. Sprains and strains	HIGH	MEDIUM			
B. Cuts and bruises	MEDIUM	MEDIUM			
C. Stress	LOW	HIGH			
D. Terrorism and the workplace	LOW	HIGH			

Figure 7.4. The four risks have been evaluated from a frequency and severity standpoint.

COMPREHENSIVE RISK MANAGEMENT PLAN

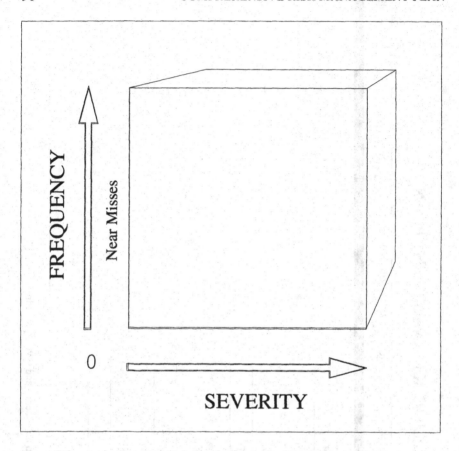

Figure 7.5. A visual tool to use during the risk evaluation process.

2. Because the scales are descriptive, not numerical, there is not a true graphing of data. The frequency scale ranges from low to high, and there are no numbers assigned to it. The same is true of severity, which can be determined in several different ways, as described earlier in this chapter. Therefore, the scale on the axis is not in number of dollars but, again, ranges only from low to high.

Experience indicates that individuals who use this system initially are confused because there is no numerical scale on the chart. However, they quickly develop a good feel for where on the chart their identified risks fit. On the severity scale, two risks that are both rated high may be charted differently based on the judgment of the individual who carries out the exercise. Remember also that there is no right or wrong answer, so little is required in terms of justifying the chart entries.

Risk Evaluation

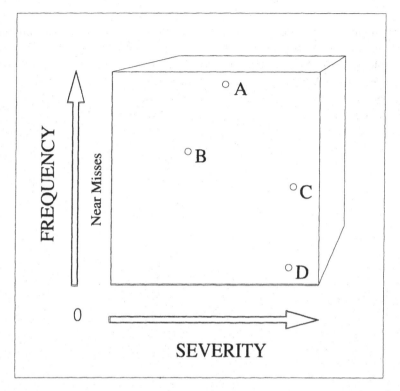

Figure 7.6. The four risks from the Sample Risk Management Plan are plotted.

Conclusion

The risks that were identified and recorded in the first step have now been evaluated. Two primary evaluation measures are used: frequency and severity. Frequency is the number of incidents that may occur as a result of a particular risk. Severity is a measure of how serious any of the incidents may be.

There is no numerical scale used for evaluating frequency and severity. Judgment is important when making these determinations, especially if they are accomplished with some basic knowledge of risks and risk management. Some guidance for evaluating these two measures is included in the body of the chapter.

An effective means of recording the results of the evaluation step are the diagrams shown in Figures 7.5 and 7.6. The primary purpose of this type of diagram is to provide a visible indicator to be used for establishing priorities. We discuss some methods of establishing priorities in the next chapter, and we shall refer frequently to this diagram. It has often been said that a picture is worth a thousand words, and the same applies to such a diagram. It can be especially useful for demonstrating what is being accomplished to those who are unfamiliar with the risk management program.

PROFILE
Chapter 8: Establishing Priorities

MAJOR GOAL:

To be able to establish priorities so that control techniques can be applied

KEY POINTS:

- Understand that there are several factors to consider when determining priorities, including:
    - Overall cost benefit
    - Cost of insurance
    - Cost of control measures
    - Ease of implementation
    - Time required for implementation
    - Estimated time to see positive results
    - Estimated effectiveness of control measure

- Understand that to establish priorities, one must consider all analysis factors, using the proper balance.

- Continue to build the risk management plan by listing the priority for each risk.

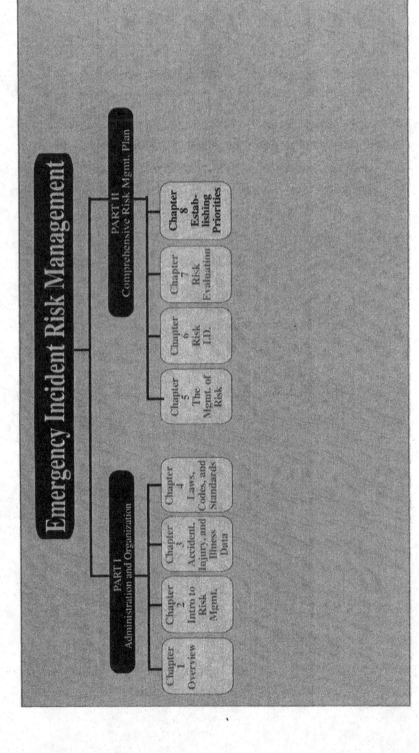

Chapter 8

Establishing Priorities

INTRODUCTION

The first step in the risk management program was to identify the risks to the organization. Secondly, the identified risks were evaluated from a frequency and severity standpoint, with the results combined and charted. Now, the information collected from steps 1 and 2 will be analyzed so that an action list can be initiated.

Now that the foundation work is completed, it is important to establish priorities so that the entity's energy and resources will be directed where they will have the greatest impact. It may be easy for all to agree on where they are headed, but the route will be in question until priorities are established and agreed upon.

The easiest and quickest way to view and analyze the information is to review the frequency/severity diagram developed in Chapter 7. That diagram is presented again in Figure 8.3, but it has been divided into quadrants numbered 1 through 4.

Quadrant 2, in the upper right, contains the risks that have been determined to be of both high frequency and high severity. In our sample risk management plan, risk A, which is strains and sprains, is located in this quadrant. Few would disagree that risks in this quadrant should be located high on the priority list.

Quadrant 3, in the lower left, is for the risks that are of both low frequency and low severity. Again, agreement is easily reached that these risks will likely be addressed last. The events involved don't happen very often, and, when they do, their impact on the organization is minimal. Our sample plan has no risks in this quadrant, but an example of a low-frequency, low-severity risk is the risk to a

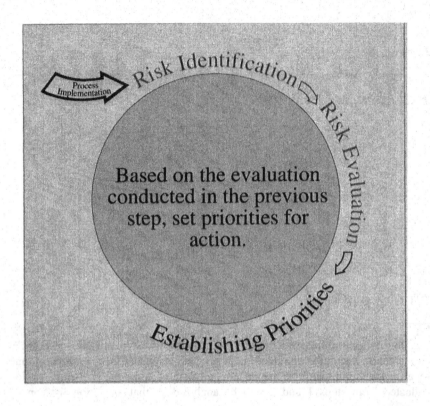

Figure 8.2. Priorities for action can be determined once the risks have been evaluated.

firefighter who packs hose without gloves and may suffer a scratch as a result of a burr on one of the hose couplings.

Quadrants 1 and 4, in the upper left and lower right, respectively, are similar. Quadrant 1 includes those risks that are of high frequency but low severity. In our sample, risk B, cuts and bruises, is located in quadrant 1. Quadrant 4 is for the risks that are of the opposite type: severe, but infrequent. Stress-related risks and terrorism-and-the-workplace risks are in this quadrant.

Of the risks contained in quadrants 1 and 4, which would be considered more important? Is it better to address the risks that happen often but that are not that serious, or to tackle the severe ones first? That is a question that has plagued risk managers for a long time, and for which there is no easy answer. However, we shall explore some factors to be considered when establishing the action list, and these same factors will help in answering the above question.

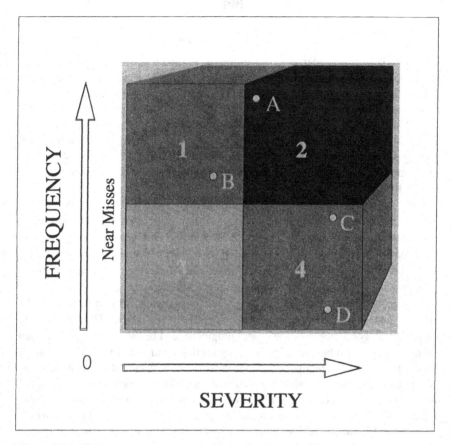

Figure 8.3. Further use of the diagram to aid with the determination of priorities.

ANALYSIS CONSIDERATIONS

There are several factors that can and should be utilized to help determine a reasonable, effective order for the action list. We shall review each individually, but they are typically used *in toto* to make an informed decision. These factors are also sometimes interrelated, so it may be impossible to act on one without impacting another. The key is to keep the end in mind, not the process. Concentrate on the fact that priorities need to be established to control accidents and injuries, not on executing an effective analysis process.

Analysis Factor 1: Cost/Benefit

What will it cost to implement control measures, and what will the return be? In most organizations the almighty dollar rules, and any expenditure that is not deemed to have a benefit that outweighs the cost will likely be denied.

It is important to remember that although the topic is personnel health and safety, the dollars that may have to be spent are real. In today's economy, it is not enough to appeal to a decision maker's moral sense of what is or ought to be right. That individual must answer to others, whether they be taxpayers or a board of directors, and something that has a perceived negative impact on the bottom line has little chance for success.

The prudent risk manager will perform basic cost–benefit analyses on various control measures so that the best set of options can be presented and defended before the decision maker(s). It may appear that this cost–benefit analysis would more likely be conducted in the next step when control measures are addressed. However, it is practically impossible to establish priorities for action without knowing the impact of control measures under consideration.

A sample cost–benefit analysis is presented in Figure 8.4. As with a real analysis, certain facts and assumptions must be identified and listed. They should be supported by good research, logic, and judgment, and be more conservative than liberal, as discussed below. The following sections will present the various steps required to conduct an analysis, and they correspond with the entries in the sample.

The first step in conducting a cost–benefit analysis is to document the costs of the risk that will be controlled. Direct costs will usually consist of medical costs and time lost from work. Indirect costs will be those incurred for lost productivity, overtime to fill vacancies due to injuries, and potentially higher insurance premiums. As outlined in Chapter 3, the best source for much of this information is the organization's past-loss records.

Once the costs have been identified, the anticipated reduction in those loss costs due to the control measure being implemented can be calculated. This is the difference between the pre-controlled loss costs and the controlled loss costs (if any).

When discussing injuries and illnesses, however, it may be difficult to prove, beyond a reasonable doubt, that control measures will result in fewer or less costly accidents or injuries. Networking, research in trade publications, or on-line inquiries may be able to provide some assistance. (Have other entities had a similar problem and adopted control measures? What has happened to their numbers since?) When conducting a cost–benefit analysis, it is wise to use extremely conservative estimates so as not to overstate a benefit.

The third step is to research and document the anticipated costs of the project. This is usually the easy part. Direct costs may consist of outlays for new equipment, personal protective gear, modifications to existing apparatus and equipment, training materials, instructors, and the like.

Establishing Priorities

Indirect costs may be numerous. They may consist of time devoted to training of personnel, overtime for additional duties, increased operating costs, higher maintenance costs, and so on. Not every control measure under consideration will have costs, and, of those that do, not all will have indirect ones. However, people who manage money or who have a say in how and where it is spent tend to be astute, so it is imperative that none of the "hidden" costs be overlooked in the research stage. It will be embarrassing if, when the results of the cost–benefit analysis are presented in hopes of gaining approval for instituting the control measure under consideration, those whose approval must be gained identify costs that were not anticipated. Remember to be prepared.

Finally, simply compare the two results to get the potential net annual benefit. If the cost of implementing the control measure is less than the anticipated annual reduction in loss costs, there is a financial benefit to the control measure. On the other hand, if the costs of the controls exceed the anticipated reduction in losses, the expenditure has not been cost-justified, and that would be considered a drawback.

Although it goes beyond the scope of this text, a factor that is frequently used when discussing the impact of future savings is net present value, or present value of future savings. It involves a method requiring several calculations, and allows financial planners to make decisions about expenditures, using the value of today's dollars. Factors such as anticipated life expectancy of the purchase, the anticipated rate of inflation, and estimated savings as a result of the expenditure are used to make the appropriate calculations.

For our purposes, the technicalities of this process are not critical. Rather, it is more important to be familiar with the terminology and to have a basic understanding of the concept. The sample cost–benefit analysis shown in Figure 8.4 does not reflect the net-present-value method, but still makes a convincing argument for the purchase of protective hoods.

Earlier, we discussed decisions based on moral rather than financial considerations. As unpleasant or as unnecessary as it may seem, even some of the moral issues can be translated into dollars. For instance, improved morale (a moral argument) frequently leads to higher productivity (improved bottom line). If employees are not getting hurt (a good, moral thing), there are no costs associated with having to deal with workplace injuries (there is a financial incentive).

The remaining analysis factors all tie back into the cost–benefit analysis, and may actually be components of it. It is still important to consider them individually, because success may hinge on a particular detail associated with one of the factors. It will also allow the cost–benefit analysis to be as sound as possible.

Cost-Benefit Analysis

INJURIES
Assume 20 burns per year at an average cost of $200 each:

    DIRECT COSTS
        Medical Costs:. . . . . ,20 / year x $200 = $4,000

    INDIRECT COSTS
        Filling Shift:. . . . . + 20 / year x $120 = $2,400

    TOTAL COST OF INJURIES (PRE-HOOD): . . . . = $6,400

Assume 15 will be prevented by the appropriate use of hoods.

    DIRECT COSTS
        Medical Costs:. . . . . 5 / year x $200 = $1,000

    INDIRECT COSTS
        Filling Shift: . . . . + 5 / year x $120 = $600

    TOTAL COST OF INJURIES (WITH HOODS): . . . = $1,600

Savings in loss costs due to the hood purchase:

    TOTAL COST OF INJURIES (PRE-HOOD):. . . . = $6,400

    TOTAL COST OF INJURIES (WITH HOODS):. . . - = $1,600

    | ANNUAL SAVINGS DUE TO HOOD USE:. . . . .= $4,800 |

CONTROL MEASURE
    DIRECT COSTS
        Cost of Hoods:. . . . . 400 x $18 / each = $7,200

    INDIRECT COSTS
        Training: . . . . Negligible; part of routine in-service training
        Maintenance:. . . . . .Negligible; cost of washing

    Life Expectancy: . . . . Two years

    | TOTAL ANNUAL COST FOR HOODS ($7,200 / 2):. . . = $3,600 |

NET ANNUAL BENEFIT = SAVINGS - COST:. . . $4,800 - $3,600 = $1,200

Figure 8.4. In this cost benefit analysis, there is a financial benefit to the purchase of hoods.

Establishing Priorities

Analysis Factor 2: Insurance Premiums

Chapter 9 will include a more in-depth discussion of the various types of insurance. When evaluating a control measure, insurance premiums may be something that are affected. Workers' compensation insurance premiums are calculated using a lengthy formula, but a key component of that formula is the entity's past worker injury claims experience. Those that have better experience will typically pay less than a similar entity that has worse experience.

That means that it makes financial sense to maintain control of the accident experience year after year. One bad year can have a dramatic impact on workers' compensation premiums for up to three years, so diligence pays off.

Other types of insurance also frequently have experience-based modification factors. An organization that has had several apparatus accidents or drivers with moving violations will likely pay higher insurance rates. Insurance is priced to cover the risks anticipated, so if the organization repeatedly demonstrates that it is "riskier" (based on its bad experience) than another, it will ultimately pay more.

Analysis Factor 3: Cost

This ties in directly with the cost–benefit analysis. The costs associated with the implementation of a control measure can have an impact on where the measure sits on the priority list. As illustrated above, costs can be both direct and indirect, and the administrator who is conducting this exercise needs to have accurate, specific information about both types.

The cost of something, whether right or wrong, moral or not, will frequently determine whether or not the thing happens. The savvy administrator will fully analyze and understand all costs associated with various control measures under consideration, and establish priorities accordingly.

For example, suppose that two risks with similar frequency/severity characteristics must be ranked. One control measure will require a capital expenditure, while the other will not. Based on knowledge of the individual or group who must approve the expenditure of funds, the administrator can present the information appropriately to most positively affect the outcome sought.

It is extremely difficult to discuss the cost factor in a vacuum, because the desired outcome of control measures frequently goes far beyond costs. The health and safety of personnel cannot be reduced to dollars, but the administrator who cannot address the cost factors will have limited effectiveness.

Analysis Factor 4: Ease of Implementation

Some things are easier, and quicker, to do than others. When establishing priorities for action, the ease of implementing the control measure should be considered. Although ease of implementation is obviously not the most important consideration, the fact that one measure is easier to implement than another may affect their relative position on the priority list.

An example of where ease of implementation may come into play would be a situation in which an organization is suffering losses as a result of apparatus backing accidents. Suppose that a proposed control measure is the installation of sensors, on the rear of the vehicles, that alert the driver to obstacles. The time and expense associated with this solution are determined to be excessive, so, instead, a policy is adopted that requires a spotter whenever the apparatus needs to be backed. It is much easier and less expensive to implement.

Analysis Factor 5: Time Required for Implementation

The best solution in the world will be ineffective if it takes too long to implement. If an entity has several risks that impact it, but the control measures for one risk will take an inordinate amount of time for implementation, that risk may be placed lower on the priority list. Why? Other efforts may suffer while all time is spent on the one exercise. Rather than making substantial progress on several fronts simultaneously, limited progress will be made on only one.

Analysis Factor 6: Estimated Time for Results

How long before one sees some results? That is not an unreasonable question. Making a long-term payoff an immediate priority can be difficult at best. Good planning, communication, and documentation become critical. Any intermediate benefits can and should be highlighted to make the upfront commitment more palatable.

Estimated time for results can be an especially difficult factor for departments that depend on elected officials for funding and support. Frequently, a politician has difficulty seeing beyond the next election, which makes any short-term results especially significant. Entities blessed with policy makers who have a long-term vision will have more success.

Analysis Factor 7: Predicted Effectiveness of the Control Measure

This factor is extremely important when conducting a cost–benefit analysis. As discussed earlier in this chapter, the predicted savings associated with the im-

Establishing Priorities

plementation of a control measure will be the component in the analysis that determines whether or not there is a benefit that outweighs cost.

The challenge of trying to predict the impact of a particular program or decision is not foreign to the fire service. Almost all agencies have some form of an ongoing fire-prevention effort. Yet the commitment of personnel and other resources to fire-prevention efforts does not guarantee success. Fires and other emergencies will likely still occur. Can one say with certainty how many fires were prevented in a particular timeframe? What cost saving is associated with those?

These same challenges arise when dealing with accidents, illnesses, and injuries, and an exmaple is the installation of intersection control devices in the response area. Such systems are not inexpensive to install. Also, personnel need to be trained in their use. However, there is no guarantee that vehicle collisions at intersections will no longer occur.

Even though results can never be guaranteed, impacts can be measured. Again, try to use the experience of others. Someone who has experience with a control measure under consideration may have data outlining any success achieved. Unless there are reasons to the contrary, success should result in other situations as well.

BALANCING THE ANALYSIS FACTORS

These analysis factors, although discussed individually here, must typically be considered in concert with each other. Their total impact needs to be measured before priorities can be established, and balance is crucial. Frequently, cost–benefit analysis will have a strong impact on the final decision, whereas ease of implementation may have a lesser one. Each is important, but should not be considered out of context with the others.

In terms of positively affecting the health and safety of firefighters, the most effective control measure is probably more effective fire-prevention efforts by citizens and use of remote-controlled robotics to handle the emergencies that still occur. Is there a net benefit to this? Probably a huge one. Are there implications in terms of ease of implementation and predicted time for seeing results? Definitely. Will this scenario occur in the near future? Probably not, although efforts to accomplish it should remain ongoing.

ESTABLISHING PRIORITIES

Now that the factors have been identified and considered, it is time finally to establish the priorities for action. As emphasized throughout this chapter, use all

SAMPLE RISK MANAGEMENT PLAN

IDENTIFICATION	FREQUENCY	SEVERITY	PRIORITY	ACTION REQUIRED / ONGOING	CONTROL MEASURES
A. Sprains and strains	HIGH	MEDIUM	HIGH		
B. Cuts and bruises	MEDIUM	MEDIUM	MEDIUM		
C. Stress	LOW	HIGH	HIGH		
D. Terrorism and the workplace	LOW	HIGH	LOW		

Figure 8.5. Priorities have been established and recorded.

the analysis factors together to determine what is most appropriate for the entity in question.

Frequently, the priorities will be evident, even to the casual observer. The risks that were identified earlier probably came as no surprise, and their corresponding controls should be relatively easy to identify as well. The judgment, knowledge, and experience of the individual(s) who will set priorities will influence the ultimate decision on what action to take first. In fact, the priority list can probably be developed in less time that it takes to read this chapter!

However, it is important to remember that such decisions are made by analyzing, even subconsciously, the factors listed above. There may be other factors as well, but, nevertheless, the priority list should not be established without adequate information.

The danger lies in action being taken without a thorough review of the appropriate information. The format of that review is unimportant: written, oral, and mental review are all effective so long as the analysis is complete.

CONCLUSION

Using numerous factors as a basis for a decision, priorities are established for the risks that were previously identified and evaluated. Because action probably cannot be taken to control all risks simultaneously, it is important to establish these priorities.

The complete priority list then becomes a risk management "to do" list. Note in Figure 8.5 that entries have been made in the priority column of the Sample Risk Management Plan. The methods available for doing what is on the "to do" list are outlined in the next chapter.

PROFILE

Chapter 9: Risk Control

MAJOR GOAL:
To determine the appropriate control technique(s) for the identified risks

KEY POINTS:

- Understand that control techniques fall into three broad categories:
    - Avoid the risk
    - Implement measures to control the risk
    - Transfer the risk

- Understand that risk control measure have to pass the "reasonableness" test.

- Understand that risk control measures can include the following:
    - Training and education
    - Standard Operating Procedures
    - Accident investigations
    - Post-incident analyses
    - Safety and health committees

- Understand that insurance is a common form of risk transfer.

- Understand that some risks will be assumed by the organization.

- List the viable control measures on the risk management plan.

- Use common sense!

Emergency Incident Risk Management

PART I
Administration and Organization

- Chapter 1: Overview
- Chapter 2: Intro to Risk Mgmt.
- Chapter 3: Accident, Injury, and Illness Data
- Chapter 4: Laws, Codes, and Standards

PART II
Comprehensive Risk Mgmt. Plan

- Chapter 5: The Mgmt. of Risk
- Chapter 6: Risk I.D.
- Chapter 7: Risk Evaluation
- Chapter 8: Establishing Priorities
- Chapter 9: Risk Control

Chapter 9

Risk Control

INTRODUCTION

Risk control is the key to the entire process of risk management. Throughout your reading of this book, your ultimate goal has been to learn to control or manage risks. Therefore, the effort expended performing the tasks outlined in the prior chapters was all prerequisite for determining the most effective way to control risks.

In determining appropriate control measures, there is a series of techniques from which to choose. These techniques, or control methods, are not unique to emergency services; they are generally accepted and universally applied by all forms of businesses. They are also inextricably linked. Typically, a combination of all the methods will be required to manage a particular risk, and the degree to which one is applied will affect the others.

We shall briefly introduce each of the control methods, then explore the strengths and weaknesses of each. An overview of the major forms of insurance is also included, because insurance is one commonly used form of controlling risk.

RISK CONTROL TECHNIQUES

These methods can also be considered a decision tree. For each entry on our priority list, we need to determine the most effective method of controlling the risk. Rather than initially jumping to what may be assumed to be the most effective control measure, it is advisable to go through each of the techniques to

110 COMPREHENSIVE RISK MANAGEMENT PLAN

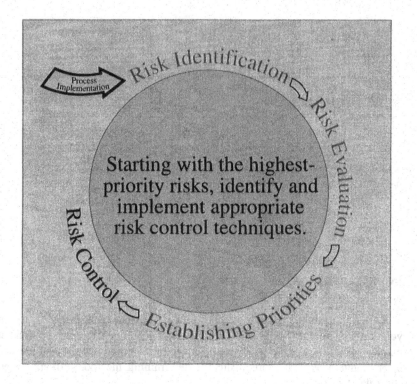

Figure 9.2. The next step is the development of an action plan for the control of the risks.

be certain that nothing is overlooked. An understanding of each of these techniques will help to clarify this premise.

The series of risk control techniques is as follows:

1. *Risk avoidance.* Simply, do not do whatever it is that creates the risk.
2. *Risk control.* This is most commonly used, and will have the greatest applicability for emergency service providers: implement measures that will help to control the frequency and severity of losses, and also reduce the overall impact of the loss on the organization.
3. *Risk transfer.* This is similar to risk avoidance: rather than avoiding the risk, transfer it to somebody else.

These three techniques, as simple as they appear, are the foundation of risk management. Any method used to control a risk will fit somewhere in this hierarchy. Volumes can be, and have been, filled with discussions of highly complex risk management principles and plans, yet all are based on these three basic premises. That is why it is so important at least to consider each of these three when deciding how best to control a risk that has been identified.

Risk Avoidance

This is the most effective means of controlling a risk; do not perform the activity that presents it. Individuals involved in emergency response activities frequently discount this method, because they are the ones called upon to respond. For example, how could a fire department avoid the risks associated with fighting fire? Simply put, it could not. However, there are certain facets of the operations of a department to which this technique will have applicability.

The strength of risk avoidance is that there are no gray areas. The activity that presents the risks is not performed. As a result, the risks no longer are a concern for the entity. Questions about the effectiveness of the solution are pointless because the risk has been avoided.

For emergency services, one of the greatest weaknesses of this risk control technique is its perceived impact on customer service. After all, someone who is called upon to respond to emergencies may have a difficult time explaining how he or she will avoid risks. However, the benefits of successfully implementing this method of controlling risk greatly outweigh the liabilities.

Risk Avoidance Example

Suppose that the risk identified is injuries to members during hose-loading operations. Some of the injuries are potentially serious, with the worst case scenario being a fatality as a result of a piece of apparatus backing over the firefighter.

The risks associated with a backing accident are easily avoided by instituting a change in the hose-loading procedure. Rather than backing the apparatus, drive it forward, straddling the hose to be packed. Members can work behind the apparatus to load the hose, and the driver need have no concerns about backing into personnel or fixed objects.

The best part about this risk avoidance control measure is that it costs nothing beyond basic training. It does require a change in habits on the part of the employees who have "always done it the old way," but the benefits greatly outweigh any potential costs associated with a change in procedure and some training.

Risk Control Measures

In the previous hose-loading example, the use of a guide while backing would be considered a risk control measure. The risk would still exist but be better controlled. The vast majority of risks that are identified will not be avoidable or transferable. As a result, control measures will have to be identified and implemented to make a potentially bad situation better.

Control measures come in a variety of forms, and there is no set list of what is, or is not, an effective control measure. Appendix A lists some of the more common risks and suggested general control measures. In fact, however, control measures are limited only by the imagination and ingenuity of those responsible for designing them. Some of the most effective control measures will never be found in a risk management text, due to their specific applicability in a particular situation.

That said, risk control measures can be identified in a variety of ways, and they will typically fall into certain broad categories, which we shall discuss. These categories should not be considered an exhaustive, limiting list. If an effective control measure is identified but cannot easily be categorized in the list, do not discard it. The goal is risk control, and the result is far more important than the process.

One test that we can apply to any risk control measure is reasonableness. We conduct operations in the real world, and our risk control measures must consider that. Control measures that fail the reasonableness test will probably not be used. Reasonableness requires acceptable answers to several questions, such as: Will the control measure work? Will members use the control measure? Will operations be affected positively or negatively? Do we have the resources to implement the control measure?

For example, suppose that a risk that has been identified is vehicle accidents during emergency response. There are several control measures to consider, one of which may state that vehicles will not respond on an emergency basis; all responses will be with the flow of traffic. Is such a measure effective at controlling the risk? Absolutely! Is it reasonable? Not at all! This control measure, although highly effective, does not even approach being reasonable, so will be immediately discarded.

Control Measures Categories

Training and Education The importance and effectiveness of training and education cannot be overemphasized, and is discussed in detail in Chapter 11. A well-trained worker will be not only more effective and efficient, but also less likely to get injured. Training allows employees to understand the proper way to perform a task, the desired result, the risks involved, and some means of controlling or eliminating the risks.

Training is not a replacement for thinking or judgment. The intent of an effective training program is not to eliminate decision making on the part of

Risk Control

the employee. Rather, it should provide that worker with a knowledge base from which to make reasonable decisions. There is no substitute for experience, especially when combined with training. If it were simply a matter of training someone to perform a task identically each time, we would hire a monkey for the task, not a person!

Training is available from many different sources. There are local, regional, state, and national training academies. Insurance companies and other industry groups usually have programs available. Probably most importantly, the organization's training officer will conduct programs.

Much of the training conducted for workers who respond to emergencies is mandated. However, whether mandated or not, training programs should have specific learning objectives outlined, be clear in their message, and measure whether or not the attendees learned anything. Attendance and a brief description of the topic, with the learning objectives listed, should be maintained to document the session.

The "Book" Standard operating procedures (SOPs), or standard operating guidelines (SOGs)? For many, many years, we have referred to them as SOPs. Now, however, there is a precedent whereby members have been penalized for not following their agency's own SOPs. In other words, if the agency has an SOP for a particular operation, and their personnel decide not to follow it, those same personnel may be penalized for their decision.

Standard operating guidelines are just what their name implies. They provide guidelines to the responding personnel to help in decision making. Because no two incidents will ever be identical, guidelines can incorporate the anticipated conditions, but still leave room for personnel to utilize judgment and experience in determining how best to handle the incident.

The decision on whether to call them SOPs or SOGs is best made locally, possibly with input from the organization's legal counsel. Regardless of what they are called, however, be sure that responding personnel have some guidance on how to manage the incident. These guidelines or procedures can be a valuable tool for controlling risks. We shall use the term *SOP* exclusively, with the understanding that agencies will decide which is more appropriate for their situation.

It has often been said that military and paramilitary organizations have a written procedure for everything, including which leg to place first in a pair of pants. These procedures are then grouped together in what most of us refer to as "the book." Hence the term "doing it by the book."

"The book" is a valuable resource for field personnel faced with a myriad of decisions that all have to be made immediately. Coupled with the extensive training that these personnel have received, the response to most incidents will be fairly standard, regardless of which personnel have responded.

Another school of thought is that "the book" was written only for those people who are not bright enough to figure out how to perform a particular task on their own. They can rely solely on "the book," and then perform adequately.

In emergency services, the most appropriate course of action is somewhere between these two philosophies, and will vary by incident. Because the book was allegedly written by people with a wealth of knowledge and experience on the topic, it is wise to heed their advice. However, because "the book" authors are probably not on the scene of the incident at this moment, the officer will have to apply his or her own judgment to make the required decisions.

What does this have to do with risk management? The SOPs outlined in "the book" describe the most effective, ultimately safest, method of handling the situation. By following "the book," responders will be less likely to get injured. Most, if not all, agencies already have SOPs in place. As a more formal risk management process is implemented, many of the existing SOPs will likely be retained. There is no need to reinvent the wheel so long as the existing SOPs remain effective.

For example, an effective SOP will describe the nature of the task, the equipment to be used, including personal protective equipment, and the steps to take to carry out the task, and will conclude with the desired result. Employees with a clear understanding of the SOP will then be more likely to utilize the protective equipment specified.

Consistent use of well-constructed SOPs will reduce the amount of freelancing at an incident. This will not only help to ensure a more successful outcome to the incident, but will also reduce the odds of injuries to responders.

Accident Investigations Emergency response agencies frequently perform incident critiques, or post-incident analyses, and these are addressed in Chapter 17. The purpose of these critiques is to learn how to be more effective the next time by reviewing what took place this time. The methods for conducting incident critiques are typically well known, and members anticipate the critique following any major incident.

This same reasoning can and should be applied to risk management. Following a serious accident or near miss, efforts can and should be made to analyze what occurred to learn how to prevent a recurrence. The approach and thought processes are almost identical to incident critiques, but are applied to the organization itself.

Accident investigations can be very basic or very complex, but they all should be considered fact-gathering exercises. The facts are available from many sources, including the victim(s), other personnel involved, witnesses, the scene of the

Risk Control

incident, tools, equipment, or other objects involved, records (e.g. weather, dispatch logs), or any other element that may be pertinent.

The goal of an accident investigation is to learn. It is not to point fingers, or place blame. The knowledge acquired following an analysis of the facts collected during the investigation can then be applied to reduce the chances of a similar incident occurring in the future. Following an accident investigation, steps, such as modifying SOPs, can and should be taken to incorporate the lessons learned.

Organizational Safety Efforts The organization somehow needs to incorporate safety into its value system. Member safety is not something that can become a priority only on the second Tuesday of each month, when the safety committee meets. Rather, it has to become a part of the fabric of the organization, interwoven into every aspect of every operation.

Many approaches are used to try and accomplish this, but they all share one theme: members taking responsibility for their own and their coworkers' health and well-being. For years, it has been assumed that organizations are responsible for this, but experience demonstrates that it truly is a shared responsibility. Although the organization can provide every opportunity and resource to establish and maintain an effective safety program, the program cannot be successful without the involvement and active concern of the people who make up the organization.

To keep safety at the forefront, and also to provide some structure to the efforts, several programs/activities can be employed. One of the most common, and most effective, is a safety committee. Called by many different names, this committee will typically have representation by employees of the major work activities of their employer, as well as from both labor and management.

The committee's charge is to serve as a safety resource for the organization and its members. Concerns about physical hazards, potentially unsafe actions or procedures, and any other factors that may impact the safety of the members of the organization are presented and discussed. Employees can use this forum to have their concerns heard, and, hopefully, acted upon.

A primary duty of the committee is to review any accidents, illnesses, injuries, or near misses that may have occurred since the last meeting. This review will include an analysis of what happened, how it happened, and, most important, how to ensure that it does not happen again. Following the review the committee will be in a position to submit recommendations that, if carried out, may prevent similar incidents in the future. These recommendations should be submitted to the individual who has the responsibility for ensuring that they are carried out.

The composition of the committee and its suggested meeting frequency are best determined locally. However, some time-tested guidelines follow.

Smaller is better. Do not make the committee so large that it is dysfunctional. Ensure that all major work activities are represented. For example, remember to include an EMS representative if the agency conducts both fire suppression and emergency medical operations.

Periodically rotate the members on a staggered basis. A fresh perspective is valuable, but continuity can be lost if all members turn over simultaneously.

Elect a chairperson, and encourage rotation of the chair periodically. Some organizations alternate between employer (officers, supervisors, managers) and employee representatives in the position.

Hold meetings frequently enough to conduct the necessary business. A larger organization with more accidents may hold a monthly meeting, whereas quarterly meetings may be adequate for others.

Have an agenda, and stick to it. Committee meetings are not the appropriate forum for "complaining sessions," and a solid agenda, if followed, will help to prevent that.

Keep everyone in the organization informed of the work of the committee, which should not be a clique with a "need to know" mentality. For example, post minutes of meetings, encourage participation by all members, and share findings of analyses. In short, communicate!

Follow up on any recommendations submitted. If they are not carried out, find out why, and take steps to try to ensure that they get carried out.

Risk Transfer

The third major risk management technique we shall address is risk transfer. This is exactly what the name suggests: transferring the risk to someone else, usually through some form of contract. Although on the surface it is easy to believe that organizations involved in the delivery of emergency services do not routinely transfer risk, they in fact do, typically through the purchase of insurance.

There are many ways to transfer risk, and some methods transfer more risk than others. If an organization contracts out its emergency medical services to a private concern, all the risks associated with those operations have been transferred to someone else. The risks still exist, so they have not been truly avoided; rather, they now "belong" to someone else on a contract basis.

Insurance, the most common form of risk transfer, actually transfers only the *financial risk* to someone else, typically an insurance company. In addition to transferring this financial risk, insurance also will help to limit the financial impact of a loss on the organization. This is an important concept, because insurance does nothing to prevent an accident from occurring. A wise risk manager will have insurance, but will also work hard to ensure that accidents do not occur in the first place. After all, the fact that there is fire insurance on a fire station does nothing to prevent the station from burning down!

Insurance is also interrelated with the other risk control techniques. Insurers look closely at the degree of risk presented by any client, so any control measures that are in place can have an impact on both the availability and price of insurance. For example, the insurer for an organization's vehicle liability coverage would be more comfortable if all apparatus drivers were licensed and trained, and in such a case the coverage may be less expensive. If there is no formal driver training program, the coverage, if offered at all, would likely be more expensive.

There are numerous varieties of insurance that can be purchased, but we shall discuss only the most common ones. Insurance is an extremely complex issue that extends far beyond the scope of this text. However, a basic understanding of the common forms of coverage is beneficial.

Workers' Compensation (WC)

This is insurance that compensates workers in the event that they become injured on the job. WC is regulated on a state-by-state basis, so there are at least 50 variations, and probably more, in existence. Each one has its own unique characteristics, definitions, limitations, and so on, but the general basis for the coverage has not changed since it was first conceived.

Workers' compensation is designed to protect both the worker and the employer. If the worker is injured or becomes ill as a result of the performance of duties for the employer, the associated medical bills will be paid by the insurance. In addition, the worker will receive indemnity benefits to make up a portion of the wages that cannot be earned during recovery and recuperation.

By accepting these benefits, the worker typically gives up all right to bring action against the employer as a result of the injury or illness. This is the characteristic of pure workers' compensation, which frequently leads to its being called the "sole remedy." No other action is allowed as a remedy to the situation.

Frequently, the indemnity benefits do not replace 100% of the injured worker's pay. The system is designed this way to provide incentive for the worker to return to work as soon as is medically possible. As conceived, WC is a bridge between the worker's original wage and return to full duty. It was never intended to be a welfare or retirement program.

Workers' compensation coverage is usually mandated by the jurisdiction in which the workers are located. Unlike the other coverages we shall address, it is not optional.

General Liability (GL)

There are many forms of liability insurance, but commercial general liability provides the broadest coverage. Organizations are advised to carry this coverage for protection from claims against it or its personnel for their actions or inactions.

For example, if a visitor to a fire station slips on a puddle of water on the apparatus floor, falls, and is injured, the general liability coverage is the one that would apply if a claim for damages is brought against the department. Similarly, some form of GL would apply if a claim is filed against the department because of allegedly excessive damage done by the department members at an incident scene (for example, venting a structure, with excessive damage, for a fire later deemed to have been minor in nature).

General liability coverage is important because it protects the financial integrity of the organization in the face of potentially large jury awards. Unlike the situation for workers' compensation, an entity can be sued, and jury trials are possible. Without the protection of the insurance policy, any awards would have to be paid out of what is likely an already strained budget.

Property Insurance

Frequently called fire insurance, property insurance would be carried on buildings and other fixed properties in which the organization has an insurable interest. These would typically include apparatus stations, offices, and storage facilities, whether owned or leased.

For example, an agency with one fire station that houses all apparatus as well as the administrative offices may also rent off-site storage space for supplies. Property insurance would be carried on the station as well as on the supplies stored in the rented space. The owner of the building in which the supplies are stored would likely carry property insurance on the building itself. The organization's "insurable interest" is only in its supplies, not the rented building, because it is owned by someone else. To take an example from a different perspective, people are barred from purchasing property insurance on a neighbor's house because they would lose nothing if the house were destroyed.

Vehicle Insurance

There are two primary forms of insurance purchased for vehicles: physical damage and liability. The physical damage coverage is for damage to the insured vehicle as a result of collision. It typically provides for repair of the damage or for reimbursement of the vehicle's actual cash value in the event it is "totalled." Vehicle liability coverage is similar to general liability coverage, but applies when the liability arises as a result of vehicle operation.

For example, in a personal injury collision that is deemed to be the fault of the apparatus driver, and in which the department vehicle is destroyed, the coverages would apply as follows.

Workers' compensation would cover the injuries to the department members. Vehicle liability would cover the claims against the department, typically for injuries and damage to property (the vehicle), filed by the other party in the collision.

Collision coverage would apply on the damage to the department's vehicle.

RISK ASSUMPTION AND RISK FINANCING

Most organizations carry insurance coverages in addition to the ones outlined above. Which ones are necessary, what limits to carry on them, and what price provides the best value are examples of issues to be addressed with the guidance of insurance professionals.

However, there will be instances in which the organization decides that it will assume some or all of the financial risk associated with a particular exposure. The reasons for this could be numerous, but price and availability of insurance are common ones. The decision that eventually leads to the assumption, or self-insurance, of a risk typically boils down to the question of how much risk will be assumed, and how the risk will be financed.

Providing emergency response services is inherently dangerous. Insurance to protect the organizations that choose to provide those services can be expensive. A valid decision-making process may lead the people making those decisions to take on some of the risk. Some of the factors usually considered include past-loss experience (has this ever happened to us before?), the likelihood that a loss will happen in the future, the quality of the ongoing risk management efforts (safety program, safety committee activities, etc.), and the financial condition of the organization.

When risk is assumed, provisions must be made to cover the costs of any claims that may occur. How best to do that should be left to financial experts, but examples include the setting aside of funds for future liabilities (claims), participation in a group self-insurance program with other, similar entities, and the purchase of a financial product, such as a bond.

CONCLUSION

This chapter has addressed the various methods of controlling risks that were previously identified and assigned priorities for action. The Sample Risk Management Plan in Figure 9.3 now includes some entries in the "Control" column. There is also a column in which a note can be made about whether the suggested

SAMPLE RISK MANAGEMENT PLAN

IDENTIFICATION	FREQUENCY	SEVERITY	PRIORITY	ACTION REQUIRED / ONGOING	CONTROL MEASURES
A. Sprains and strains	HIGH	MEDIUM	HIGH	ONGOING ONGOING ONGOING	1. Conduct periodic awareness training for all members 2. Evaluate function areas to determine location and frequency of occurrence. 3. Based upon outcome of evaluation, conduct a task analysis of identified problems.
B. Cuts and bruises	MEDIUM	MEDIUM	MEDIUM	ONGOING ONGOING	1. Review SOP on use of PPE, both emergency and non-emergency. 2. Determine whether PPE will reduce the number of incidents, based upon analysis.
C. Stress	LOW	HIGH	HIGH	ONGOING ONGOING	1. Continue health maintenance program. 2. Participate in physical fitness program.
D. Terrorism and the workplace	LOW	HIGH	LOW	ONGOING ONGOING	1. Provide awareness training for all personnel. 2. Develop policy and procedures as indicated by need.

control is already in place (*O* for *ongoing*), or whether action is required (*A* for *action*).

Although the most common risk management techniques were presented individually, a combination of some or all of them will usually be the most reasonable, effective solution. The best way to explore the options is to use them in a thought process or decision tree. Do not discount any possibility without at least considering it. A review of each step in the process may turn up a previously unconsidered solution. This examination process will quickly allow the viable solutions to be identified, and the others to be discarded. Typically, this can be done more quickly than the time it takes to describe it to someone else. The specific details of a control plan may take more time, but the decision on the best course of action is usually not a time-consuming one to make.

Above all, include a strong dose of common sense in this decision-making process. The technicalities of risk management and risk control are far less important than the desired outcome, which is fewer and less severe accidents and illnesses. This is not intended to be a mechanical, procedure-driven process; it is intended to help those who are responsible for the health and welfare of members to make sound decisions.

PROFILE

Chapter 10: Program Monitoring

MAJOR GOAL:

To understand the importance and method of monitoring the risk management program to ensure that it remains effective

KEY POINTS:

- Understand that the purpose of periodically monitoring the risk management program is to ensure that it is effectively serving the organization.

- List the indicators that can be used to evaluate effectiveness:
  - Number of losses is trending downward
  - Cost of losses is trending downward
  - Lost work time is decreasing
  - Health and safety recommendations are being completed
  - Laws, codes, and standards compliance is increasing

- Learn to conduct the evaluation a minimum of annually, and use somebody from outside the organization to do so every three years.

- Understand that there are several sources of evaluators, both internal and external.

- Understand that to be comprehensive, the evaluation should include some specific components.
  - Review of past loss experience
  - Interviews with key staff members
  - Random interviews with other staff members
  - Pertinent SOPs

- Understand that a written report that documents the findings and recommendations should be filed after it has been discussed with the appropriate personnel.

- Based upon the results of the program review, either internally or externally, make program modifications.

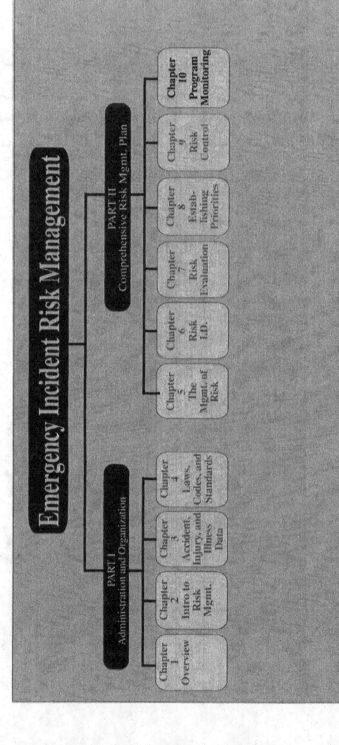

Chapter *10*

Program Monitoring

INTRODUCTION

So far our risk management program has addressed risk identification, an evaluation of the identified risks in order to establish priorities, and the formulation and implementation of effective control measures. The final step in the model is to evaluate the effectiveness of the entire program.

The purpose for doing so is to determine what is working, what is not, and what may be required to make the program better serve the organization. The monitoring can be as simple as the filing of a written report by the health and safety officer to top management, or it may be as complex as the preparation of a lengthy study and report by an outside consultant. There are several factors to consider prior to conducting this evaluation, and we shall explore them in this chapter.

PROGRAM EFFECTIVENESS

The purpose of conducting an evaluation of the risk management program is to determine whether or not it is effective. But how do we measure effectiveness? This may be a rhetorical question, but it should be carefully considered prior to undertaking a program evaluation.

Unfortunately, there is no hard and fast effectiveness scale that we can use. However, there are several indicators that can be considered. These may include the following.

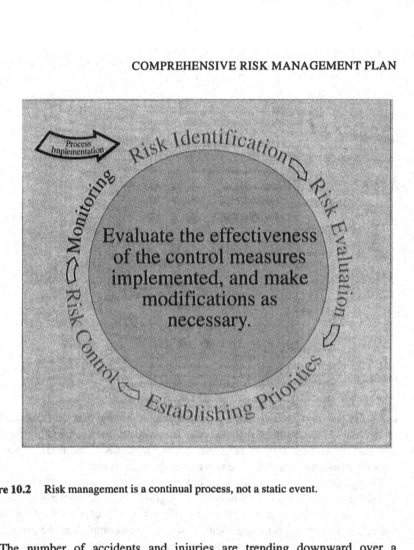

Figure 10.2 Risk management is a continual process, not a static event.

The number of accidents and injuries are trending downward over a meaningful period of time.

The cost of accidents and injuries is trending downward over that same period of time.

Time lost due to work-related injuries or illnesses is reduced.

Recommendations addressing health and safety issues are made and completed.

The organization is in, or working toward, compliance with applicable laws, codes, and standards.

All these factors are tangible, or easily measured. Other factors to consider are the less tangible ones such as productivity, employee morale, and quality of customer service. After all, an organization that conducts business efficiently and productively and that exceeds the expectations of its customers is probably also conducting it safely. The disruption caused by accidents, injuries, and illnesses detracts from the delivery of services, so anything that is done to prevent these "distractions" will pay dividends.

A final factor we shall list is judgment. Although not measurable, and probably not cited in any other risk management textbooks, the perceptions of those people responsible are important, whether they are chiefs of department or other top managers, health and safety officers, shift commanders, or line firefighters. A sense on the part of such people that something is amiss should not be overlooked. Initially, it may be difficult to pinpoint the cause or causes for concern, but the search will probably be worthwhile. Judgment is critical for success in this business, and that is as true in what may be considered administrative tasks as it is in tasks that carry life-and-death implications.

FREQUENCY OF MONITORING

The most effective monitoring is that which occurs on a continual basis. Typically, it will be informal and be conducted by every member of the organization. If there are questions or indications that a modification of the program may be required, they can be called to the attention of the individual(s) most able to make modifications, such as the health and safety officer. This is analogous to situations that involve the monitoring of health. If you exhibit the warning signs of heart trouble, you will probably see a doctor immediately and not wait for your next regularly scheduled medical evaluation.

There will be more-formal program evaluations conducted at a specified frequency. NFPA 1500 recommends that an annual internal evaluation be conducted. That is, the organization is to examine its own program and make modifications as necessary. An advantage of having someone from within the organization conduct the evaluation is that he or she will be thoroughly familiar with the operations, facilities, and personnel that the program addresses. A disadvantage is that this same familiarity removes some degree of objectivity. In some cases, the loss of objectivity can be exhibited as apathy on the part of the reviewer, causing some of the supposed benefits of the evaluation to be lost.

NFPA 1500 goes on to recommend an evaluation by somebody from outside the organization a minimum of every three years. The purpose of having someone from the outside conduct an evaluation is to reduce the chances that members of the organization will overlook something that may be readily apparent to someone else. It is easy to see the same situation on a daily basis and make assumptions about what is, or is not, occurring. This evaluation by an independent third party can be both an objective review of the organization's risk management plan and a valuable learning experience for those who routinely conduct the in-house evaluation.

In addition to the continuous review, and the scheduled ones suggested by NFPA 1500, there are other indicators that should trigger an evaluation. These indicators can be considered warning signs that there is a breakdown in the program that may require immediate maintenance. Some examples of these indicators are listed below. However, the list is not all-inclusive. It is incumbent upon the members of the organization to be alert to these and other warning signs so that unnecessary injuries and illnesses do not occur. The warning signs include:

A death or serious injury
An incident or series of incidents that requires action to prevent a recurrence
Post-incident analyses (PIA) that identify the need for program modifications
A near miss with potential for serious consequences

WHO CONDUCTS THE EVALUATION?

As described above, both internal and external sources will be used to conduct the program evaluations, and examples of characteristics of each are listed below. The decision of whom to use will depend heavily on local circumstances, but there are several factors/qualifications to consider. These are not listed in any order of priority, because a varying combination of all of them will determine who is most qualified for any particular assignment.

Interest This is especially important for personnel from within the organization who may be involved. Regardless of other qualifications, a lack of interest in the subject matter will likely result in an inadequate review.

Expertise Knowledge of the risk management process, especially as it impacts organizations involved in emergency services, is a bonus. Formal training on the subject matter is available, but internally this expertise may be acquired through participation in the safety and health committee, for example.

Experience Frequently, experience will contribute to improved expertise, but the two are separate attributes. Someone who has successfully completed other program evaluations may be able to provide experience-based guidance for others. The best experience is both wide and deep. Someone who has conducted evaluations for several different organizations of varying characteristics (e.g. size, operations) will have a broader perspective than an individual who has conducted the same number of evaluations, but for only one entity.

Analytical ability Portions of the evaluation may require the review of accident and injury data or the compilation of dollar-loss information. The ability to list and sort the numbers is not nearly so important as the ability to understand and analyze them. Data can provide vital information.

Communication skills This is important in two ways. First, a thorough evaluation may require interviews with key personnel in the organization. An outside consultant will probably want to discuss the risk management program with top management, as well as with the health and safety officer, for example. The ability to "speak the language" of the fire department will make this a more productive exercise. Second, the report on the results of the evaluation needs to be clear, concise, understandable, and usable. It may well become the instrument used to make major modifications to the program, so the included information must be clearly communicated.

Internal Evaluators

Frequently, quality evaluators can be found within the organization. For example:

Chief of department, or other top manager/administrator It is unlikely that this individual will routinely conduct the evaluation. However, the act of conducting the evaluation presents a good opportunity both to demonstrate a commitment to the health, safety, and well-being of members and to keep current on the details of the plan.

Health and safety officer The person in this position is the most likely candidate for conducting the majority of the evaluations. The HSO has an intimate knowledge of the organization's health and safety programs, knows the history of accidents, illnesses, injuries, and near misses, knows the people, and understands the intricacies of the risk management program.

Health and safety committee A review of the risk management program may provide valuable experience for the committee. Doing anything by committee is usually less focused and more time consuming, so it is not advisable to have the committee conduct all evaluations. This is not to say that the committee should not be kept apprised of the status of the program. Their input and guidance can be valuable, but the mechanics of conducting the review are better left to an individual.

Safety division/department head In organizations that have a safety division or a department head responsible for the safety function, the evaluation would likely occur there. In much the same way that the health and safety officer understands the most about the risk management program, this division should also possess the requisite background information.

External Evaluators

There are numerous sources to explore in searching for an evaluator from outside the department. Some will charge for their service, whereas others will not. The point is that this does not necessarily have to be an expensive proposition. Examples of some sources of external evaluators follow.

Chief or other top manager of a neighboring department As described above, for any top administrator, this may not be a routine source for assistance. However, chiefs may periodically want to compare and agree to evaluate each other's programs.

Health and safety officer from a nearby department Health and safety officers frequently have the background and skills necessary for this exercise, and the perspective from an HSO from outside the department will be valuable.

Insurance carrier Loss control or loss prevention consultants or engineers, as described in Chapter 3, are trained and available to provide guidance for the prevention of accidents. Program review and evaluation are services that should be offered, and that can and should be requested.

Municipal risk manager An evaluation conducted by someone in this position will most likely be directed at the true risk management components of the program rather than department specifics. In other words, an organization will be more likely to learn from a municipal risk manager whether it has effectively addressed the basic steps of risk management than it is to learn whether its risk identification list is complete. As with some of the other sources of evaluators, this may be valuable feedback periodically.

Consultant Consultants are available to conduct this type of review. Prior to enlisting the services of a consultant, it is important to agree on both the purpose and scope of the project. Keeping the consultant's efforts focused will ultimately be more cost effective. Quality recommendations for improving a risk management program can be generated by a consultant who has a familiarity with emergency service providers and their operations.

EVALUATION METHODOLOGY

Everyone will have a different method for evaluating a risk management program. No system is better or worse than another, but it is important for the evaluation to be comprehensive. In order to ensure that it is comprehensive the following steps, at a minimum, should be included.

A review of past loss experience for the organization This would include all injuries, illnesses, and exposures to personnel, apparatus accidents, losses of or damage to tools and equipment, etc. It is recommended that this review encompass not only the time period under review, but several prior years as well, to ensure that no trends are overlooked.

Interviews with key staff members This would typically include the administrator who has responsibility for the risk management program, the health and safety officer, incident safety officers, and any others who have some level of responsibility for, or involvement with, the risk management program. The purpose of these interviews is to learn more about the existing program, its current status, any planned revisions (and why they are planned), and any other factors that may affect the program. An example of information that could come to light during one of these interviews is information that the City Council has made it clear that budgets for the next fiscal year will be level-funded, and that any planned capital expenditures will likely be delayed for at least a year. Although this information will not impact the evaluation of the current risk management program, it may affect the timeline for the completion of any recommendations that are submitted following the evaluation.

Interviews with other staff members chosen at random Sometimes, the most is learned from these interviews. Regardless of how allegedly well designed, administered, and executed a plan is, it is only as effective as those who must operate under it perceive it to be. These random interviews, which should include personnel from different divisions (e.g. communications, operations, training, maintenance) will provide information on how well understood the plan is, whether it reaches the people it is designed to protect, and whether the goals of the plan are being realized and achieved. These interviews need not be formal. A simple discussion on the apparatus floor will usually suffice. In fact, once initiated, these discussions can take on a life of their own and provide reams of useful information on a variety of topics.

A review of pertinent SOPs Included would be those that have a direct impact on health, safety, or other risk management function, and any that the evaluator believes should be reviewed as a result of the review of past losses. For example, if several apparatus incidents were noted in the review of losses, the SOPs addressing apparatus operation and operator training may be included in the review process. More difficult to identify, but equally important, are the SOPs that do not exist but that should. This information will appear again when modifications to the program are recommended, but during the review of SOPs the evaluator should be mindful of omissions. It is far easier only to review those that already exist, but a quality evaluator will be also able to identify and recommend solutions for the omissions.

In smaller organizations, much of this information can be gathered and reviewed at one sitting. With adequate preparation and planning, a typical review might proceed as follows.

The individual conducting the evaluation will review the loss records prior to making the site visit.
Interviews with key personnel, which were scheduled in advance, are held. Questions about the loss data are addressed.
Interviews with on-duty personnel are conducted at a time that is most convenient for them.
Between interviews, SOPs are reviewed.
A closing conference is held with the appropriate personnel, usually the health and safety officer, to address any remaining issues.

After the day spent on site, the evaluator prepares the report and makes any recommendations needed. A follow-up meeting is held to discuss the findings and to attempt to reach agreement among the parties on the validity of the evaluator's recommendations. Agreement is important, because it increases the likelihood that the recommendations will be carried out. Whether agreement is reached or not, the completed report is submitted following the discussion.

In larger organizations, the review may span several days on site. Adequate planning and preparation remain extremely important, and the evaluator can again review much of the necessary information, such as past losses, and even SOPs, in advance. The time on site will be longer simply because the organization is more complex, with more departments/divisions, more personnel who need to be interviewed, and, as a result of these same factors, a risk management program that is more involved.

Regardless of the time spent in advance and on site, however, the actions of the evaluator following an analysis of all information reviewed remain the same. Findings are recorded and recommendations are generated, but they are not submitted until they have been discussed with the appropriate parties.

RESULTS OF EVALUATION

Once the reviews and interviews are completed, the evaluator will process the information. The intent is to have a clear understanding of the goals and objectives of the risk management program, and to make a determination on whether those goals and objectives are appropriate and whether they are being met. If any of the goals and objectives are not being met, or if other areas needing improvement are identified, recommendations can be formulated to modify the program to help get it back on track.

Findings, or conclusions, will fall into two basic categories. The first, and probably the more important, category includes the positive factors that were identified. After all, the overall purpose of the exercise is to make the risk management process proactive in addressing health and safety issues, and the vast majority of organizations are already doing something right. The most effective way to ensure that positive behavior continues is to recognize and support it. This positive behavior will lead to positive results, so, when identified, the evaluator needs to note such behavior.

The other category includes the areas that may need attention. Some refer to these as the negative components of the program, but a more constructive approach is to consider them as opportunities for making improvements in the program. These are typically easier to identify and quantify than the positive aspects, but it is not the sole purpose of the evaluation to identify problems.

A part of this category will be recommendations for improvement. As was stated above, the purpose of this is not simply to identify problems. Once a problem is identified, it becomes more important to identify alternatives for solutions, and at that point the individual who conducts the evaluation can provide the best service. If the evaluator is someone from outside the organization, the recommendations can be based on experience in dealing with other, similar organizations, and can include guidance of a sort that has worked for others.

Recommendations, to be effective, should be worded in such a way that the recipient will understand what is being recommended and why it is being recommended, and will know when the recommendation has been satisfactorily completed. It is also appropriate to include assignments (who will do what) and target dates for completion, especially if there are several components to the same recommendation.

All these findings (both positive and negative) and recommendations need to be documented in a written report, regardless of who conducts the evaluation. The report, although probably addressed to the fire chief or other top administrator, should also be made available to the health and safety officer, and the occupational safety and health committee. This is especially true if those individuals or groups will have some responsibility for completing any recommendations included.

Conclusion

Conduction of a periodic evaluation of the risk management process is the last step before the process begins anew. It is important to remember that the purpose of the evaluation is to identify any weaknesses in the program so that actions to modify and improve it can be taken. Recommendations that have been generated by the evaluator and that have the full understanding and agreement of the key people in the organization are much more likely to be completed. Reports with recommendations that surprise or, worse, embarrass the recipient will not result in any positive action being taken.

In this section, we have discussed the classic risk management model. It is as much a thought process or decision making tool as it is a model. The advantage of understanding this approach to managing risk is its applicability to the risks presented by emergency incidents. Using Parts I and II of the book as a foundation, Part III will now address emergency incident risk management.

PROFILE
Chapter 11: Training of Personnel

MAJOR GOAL:

To ensure that all organizational personnel are adequately and properly trained and maintain certification commensurate with their level of expertise

KEY POINTS:

- Identify and use nationally recognized training standards and competencies.

- Develop, implement, and utilize a department-wide training matrix.

- Have a recordkeeping system that tracks all mandated training (e.g. OSHA) and certifications (e.g. EMT, Paramedic).

- Incorporate risk management and safety into all aspects of the training program.

- Ensure that certified training is provided, conducted, and completed prior to any participation in emergency operations for all new personnel.

- Remember that "the way we practice is the way we play."

- Recognize that training and education comprise a very proactive program that affects all aspects of the organization.

- Understand the link between effective training and personnel health and safety.

Emergency Incident Risk Management

PART I
Administration and Organization

- **Chapter 1** Overview
- **Chapter 2** Intro to Risk Mgmt.
- **Chapter 3** Accident, Injury, and Illness Data
- **Chapter 4** Laws, Codes, and Standards

PART II
Comprehensive Risk Mgmt. Plan

- **Chapter 5** The Mgmt. of Risk
- **Chapter 6** Risk I.D.
- **Chapter 7** Risk Evaluation
- **Chapter 8** Establishing Priorities
- **Chapter 9** Risk Control
- **Chapter 10** Program Monitoring
- **Chapter 11** Training of Personnel

Chapter *11*

Training of Personnel

INTRODUCTION

Why is it important for an organization to ensure proper mandated training and recertification of this training? Members of an organization who provide an emergency service to a community need to be trained adequately to handle a variety of emergency incidents. In the past, on-the-job training was the primary way of gaining this experience. Experience is an excellent teacher but can be an expensive one. Personnel must be properly trained to ensure they know how to function safely and effectively during an emergency. Allowing untrained personnel to operate at an emergency incident is irresponsible, reckless, and often costly.

The goal of any training program is to provide the necessary instruction and education to properly qualify personnel to perform at a particular level of competency and in a safe and efficient manner. In the past, training programs consisted of minimal classroom instruction, "hands-on" experience, and a great deal of on-the-job training. The need for and importance of an effective training and education program have been identified, based upon nationally recognized standards and competencies as established by various training agencies. As identified in the risk management process, training is an essential and fundamental control measure. As safety and health becomes an organizational value, training and education programs must continue to enhance safety and this interaction must exist between both.

As personnel join a fire department, EMS organization, or industrial brigade, usually the first contact they have is through the training program. This is the point at which the organization has the opportunity to introduce and instill its safety and health philosophy. Regardless of whether personnel have initial knowledge or understanding of the job at this time, it is critical that the safety and health groundwork be established. This safety and health foundation will, hopefully, continue to strengthen throughout their careers with the organization. The organization must take the opportunity to develop positive attitudes toward safety and health. If that is not done during the initial training, such as in recruit school, it becomes more difficult to initiate later. Because it is human nature to refuse or reject change, an organization must establish its safety and health goals and use them as a basis for training, whether it be for recruits, company in-service, or officers.

The fire service has learned painful lessons as a result of ineffective and inadequate training. In the past, personnel received minimal training and then were sent to a company for on-the-job training. The problem with the on-the-job training was that new personnel learned unsafe practices from the personnel teaching them. Organizations can now utilize a variety of training options to ensure that personnel are properly trained and certified. Personnel are trained to perform to their level of competency safely and effectively.

TRAINING—A VITAL COMPONENT OF PRE-EMERGENCY RISK MANAGEMENT

As part of the comprehensive risk management process, pre-emergency risk management issues must be identified, as described in Chapter 12. An organization must ensure that personnel have been properly trained and certified before allowing them to respond to emergencies. Whether the organization itself provides the training or utilizes outside instructors, the organization is responsible to ensure that personnel are qualified to perform the tasks to which they are assigned.

In order to ensure that personnel are trained to a particular level of competency, competent standards must be utilized. There are several means by which to accomplish this objective. The most commonly used training standards are developed and issued by the National Fire Protection Association (NFPA). Some state training agencies have developed and utilized their own certifications. In order to meet the intent of NFPA 1500, *Standard on Fire Department Occupational Safety and Health Program*, they must provide proof of equivalency. This is defined in Section 1-4, "Equivalency," in NFPA 1500. If organizations do not utilize NFPA training standards, the authority having jurisdiction (e.g., state training director, fire chief, board of directors, personnel department) must certify that this training is equivalent to the appropriate NFPA standard. An equivalency statement must be given in which the authority having jurisdiction ensures that the training, education, competency, and

Training of Personnel

Figure 11.2. Fire and Emergency Medical Service personnel receive technical rescue practical training under controlled conditions (photo by Murrey Loflin).

safety that are being provided parallel the training, education, and competency that are commensurate with the duties and responsibilities of the members, regardless of the training standards utilized by the authority. Equivalency will not jeopardize in any manner the competency and safety of members in lieu of compliance. The following is an examination of professional qualification standards for various positions within a fire and/or EMS organization:

Firefighter Provides competency levels for personnel who participate in interior structural firefighting. They must meet the objectives for Level I and Level II in NFPA 1001, *Standard on Fire Fighter Professional Qualifications*. This standard addresses topics such as firefighter safety, fire behavior, portable fire extinguishers, building construction, ground ladders, ventilation, hose practices, and forcible entry. The intent of NFPA 1001 is that personnel trained to this degree can operate at an emergency scene under the direct supervision of an officer or senior firefighter.

Driver/operator Provides the minimum levels of training for personnel assigned to drive and operate fire apparatus. The organization must provide this training before allowing personnel to drive fire apparatus under emergency and nonemergency conditions. Most entry-level personnel have never driven a vehicle the weight and size of fire apparatus. NFPA 1002, *Standard for Fire Apparatus Driver/Operator Professional Qualifications* provides requirements for personnel who drive and operate fire apparatus under emergency conditions. All personnel who operate vehicles, regardless of type, should complete the National Safety Council "Defensive Driving Course."

Emergency medical technician, cardiac technician, paramedic Personnel who provide emergency medical care must be properly trained and certified prior to service delivery. The qualification and recertification programs vary from state to state, but an organization must ensure that initial training, continuing education, and recertification are documented, in the event of a liability issue.

Officers Personnel in supervisory positions should complete jurisdictional training for supervisors as well as NFPA 1021, *Standard for Fire Officer Professional Qualifications*, or an equivalent program. This standard covers such topics as safety, fire prevention programs, recordkeeping, emergency operations, and fundamental supervisory procedures. As a member is promoted through the ranks, the level of fire officer certification will be dictated. Another program for consideration is the National Fire Academy's "Executive Fire Officer" program, which is for mid-management to upper-level-management personnel.

TRAINING AS A RISK CONTROL TECHNIQUE

As was discussed in the classic risk management model in Chapter 9, there are three basic risk control techniques. We shall discuss risk avoidance, risk control, and risk transfer, and determine the applicability for each in the training function.

Risk Avoidance

Suppose that the XYZ Fire Department has personnel trained to the hazardous materials awareness level per the requirements of 29 CFR 1910.120, *Hazardous Waste Operations and Emergency Response*. Personnel are attempting to stop a tank truck from leaking an unknown product. Through effective training and education, the first arriving officer determines they have neither the resources nor the expertise to handle this scenario and must call for assistance from qualified personnel while waiting from a safe distance. Through effective risk assessment, the officer identified the magnitude of this incident. The officer avoided jeopardizing the safety and welfare of personnel from the XYZ Fire department.

Risk Control

Simply stated, effective training for personnel is a form of risk control. Well-trained personnel function more safely than poorly or inadequately trained personnel. The use of standard operating procedures (SOPs) ensures that personnel are provided the necessary direction to conduct an operation or activity, whether it be emergency or nonemergency.

Another example of risk control is the development, implementation, and management of a safety and health program. An organization that does so is taking steps to control risks through written procedures outlining detailed requirements for compliance with recognized practices and principles.

Risk Transfer

Risk transfer is described as the transfer of risk to another person, company, or group, or the obtaining of insurance against the risk. Based upon these descriptions, training and risk transfer would rarely make a viable combination.

Risk control techniques play a valuable role in the training function. An organization must effectively utilize this technique as part of the overall risk management process.

ACCIDENT PREVENTION AND TRAINING

The health and safety officer is responsible for developing and implementing an accident prevention program within the structure of the organization's training program. This program can function in a variety of ways and can be taught by the health and safety officer or other trained staff.

First, such a program should be included as part of the recruit training program. Personnel will be instructed on the proper methods of completing tasks in a safe and effective manner. Personnel must understand that they are a valuable asset to the organization. If an injury occurs to a member of the department, regardless of its severity, the organization may lose the services of this member. Also, an expenditure such as workers' compensation is incurred. This does not mean that the department is not concerned about the member's welfare, but the department does not want the member to suffer an occupational injury or illness.

Second, as personnel are promoted into supervisory positions, the accident prevention program should be revisited and the supervisor's role should be included in this process. The supervisor is a key player in the safety and health program. He or she must be part of the safety team in selling the accident prevention program to personnel. Without his or her support and input, the results will be minimal at best. It is impossible for the safety and health officer to be at all fire department operations. It is therefore of great importance that company officers be able to assume this responsibility.

Based upon studies conducted over the years by safety professionals, the majority of accidents that occur are caused by human error. These studies have shown that the percentage of accidents caused by human error can be 85% or higher. Equipment failure accounts for only about 5% of accidents and injuries. It does not take a rocket scientist to determine which percentage to focus the training needs upon.

In the past, the fire service did little to protect itself from accidents and injuries. Through the development and emphasis on safety and health during the past 10–15 years, this philosophy is slowly changing. Statistics show that a reduction in accidents and injuries reduces lost work time, reduces workers' compensation costs, decreases liability and insurance costs, and, more importantly, leads to a healthy workforce. The accident prevention program is a tool to enhance the department's occupational safety and health program.

LIVE TRAINING EVOLUTIONS

An excellent example of implementing risk management is the development of NFPA 1403, *Standard on Live Fire Training Evolutions in Structures*. This standard has been one of the most important factors in reducing firefighter fatalities and injuries in nonemergency environments. Prior to the development of this standard, there were no written guidelines on how to conduct the evolutions in a safe and effective manner. There was little or no regard for the safety of personnel involved in these operations. After several significant incidents that included firefighter fatalities and/serious injuries, the fire service began to realize the need for guidelines when conducting live fire training. The safety problems included use of flammable materials to burn, poor water supply and back-up lines, lack of proper personal protective equipment, lack of pre-planning or escape routes, and lack of an incident management system.

The advent of NFPA 1403 provided strict requirements for conducting live fire training in burn buildings or acquired structures. NFPA 1403 addresses issues such as:

Figure 11.3. Firefighter recruits prepare to participate in a live fire training evolution (photo by Martin Grube).

    Safety, including use of a safety officer
    Environmental impact
    Disconnecting utilities
    Building construction and condition
    Use of proper personal protective equipment
    Proper water supply and hose lines, including back-up lines
    Pre-planning with students about escape routes
    Use of incident management system
    Instructor/student ratio
    Emergency medical care on the scene

Without these guidelines in place, the organization can greatly increase its liability and jeopardizes the safety and welfare of personnel.

Another example of live training evolutions includes mass casualty drills and hazardous-materials incident drills. Mass casualty drills could involve a variety of situations that could involve a significant number of victims. Hazardous-materials incidents could simulate product release of a hazardous material at a fixed site or during transportation of product. The intent of these drills is to test emergency response agencies' capabilities to mitigate or control the situation in the best possible manner.

An important part of this evolution is to test the department's or agency's incident management system. Mass casualty incidents are very difficult on an organization, so member health and safety is a very significant priority. Hazardous-materials mitigation can be a lengthy process that requires a commitment of time, resources, and personnel. Examples of goals that would need to be addressed from a risk management standpoint in conducting a live training evolution are:

Incident scene safety
Member welfare
Rehabilitation (due to the length and nature of the incident)
Accountability

MANDATED TRAINING

Why do we train and educate our personnel before allowing them to function in emergency incidents? Are we mandated to do it by federal or state law (e.g., Occupational Safety and Health Administration), jurisdictional regulations (e.g., state EMS training agency), or organizational regulations? With more emphasis being placed on health and safety in addition to the evolution of NFPA 1500, *Standard on Fire Department Occupational Safety and Health Program*, fire and EMS agencies are more aware of applicable laws and standards. Is a 30-member volunteer fire department located in a state that is under Federal OSHA, mandated to comply with the requirements of 29 CFR 1910.1030, *Standard on Bloodborne Pathogens—The Final Rule*? Technically, it is not so required, because governmental agencies such as municipalities and volunteer agencies are exempt from the requirements under the OSHA provisions. However, we have to explore all the options before closing the book on this issue. An administrator of this volunteer department must look at the legal ramifications of failure to comply with a nationally recognized regulation. More importantly, the administrator must evaluate the risks to the members of the organization who are providing the service. The significance of the infection control issue in this country today dramatically impacts this decision. The administrator must be able to legally justify the decision for noncompliance with the *Bloodborne Pathogens* regulation, which may be impossible to do. The options for compliance would be:

Proper level of emergency care or standard level of emergency medical care provided to external customers
Humane and moral obligations
Safety, health, and welfare of the department members

Training of Personnel 145

Reduction of short-term and long-term health maintenance costs
Reduced liability against the organization
Reduced insurance and risk management costs for the organization

From an OSHA regulation standpoint, it is imperative that an organization ensure compliance with the applicable standards. What are the mandated regulations that affect us in our daily operations? The following is an example of general industry standards that require training for members of an emergency response organization:

29 CFR 1910.120, *Hazard Waste Operations and Emergency Response*
29 CFR 1910.134, *Respiratory Protection*
29 CFR 1910.146, *Permit-Required Confined Space*
29 CFR 1910.156, *Industrial Fire Brigades*
29 CFR 1910.1030, *Bloodborne Pathogens*

For example, if the ABC Volunteer Fire Department lacks the expertise and the resources to provide confined-space rescue, do the members need to be trained or educated on 1910.146, *Permit-Required Confined Space*? The answer is *yes*! The members need to be trained to an awareness level sufficient to ensure they know the limitations if they respond to an incident involving a confined-space entry.

NFPA 1500

Prior to the development of NFPA 1500, *Standard on Fire Department Occupational Safety and Health Program*, there was no written safety and health program solely for the fire service. NFPA 1500 provides a comprehensive safety and health program that incorporates into a single document all functional areas affecting a fire department. Topics include training and education, apparatus and equipment, personal protective equipment, incident scene safety, facility safety, health maintenance, and member assistance programs. This document can serve as an excellent resource for EMS organizations that are developing and implementing a safety and health program. Obviously, not all requirements will pertain to EMS, but the majority of requirements will impact the organization. It is very simple to change the term "fire department" to "EMS" in these requirements. This will provide a solid basis for developing a risk management program.

The National Fire Protection Association (NFPA) provides a variety of certifications for firefighters, officers, and other specialized functions. As discussed earlier in this chapter, an organization must ensure that personnel are trained to a recognized standard of competency or the equivalent. This includes:

NFPA 1001, Standard for Fire Fighter Professional Qualifications
NFPA 1002, Standard for Fire Apparatus Driver/Operator Professional Qualifications
NFPA 1003, Standard for Airport Fire Fighter Professional Qualifications
NFPA 1021, Standard for Fire Officer Professional Qualifications
NFPA 1031, Standard for Fire Inspector Professional Qualifications
NFPA 1041, Standard for Fire Service Instructor Professional Qualifications

For the first time, the fire service has seen proactive changes based on the development and implementation of health and safety programs. Strangely enough, this paradigm has met serious opposition from many facets of the fire service. Steeped in tradition, the fire service has been slow to accept this change.

This is why, from a risk management perspective, organizations must start at the beginning and evaluate their effectiveness to respond to and control emergency incidents. There is no better place to start than with the training program for personnel. There, safety and welfare are paramount. An organization must develop a plan to ensure the proper or necessary certification for personnel.

Each position that a department utilizes must have a training matrix that defines the minimum qualifications for that position. The training matrix defines qualifications for each position and recertification as applicable. A sample matrix is shown in Figure 11.4.

From the standpoint of our customers, we must provide a competent delivery of service within our community. The old adage is: "We play the way we train." This saying has a great deal of merit. If we do not train or if training is done improperly, we cannot expect the incident scene to be any better. With regard to emergency scene risk management, an organization or department must have standard operating procedures in place to enhance and augment the training program. All of us operate somewhat differently, based upon local conditions, resources, and restrictions, but we have to conduct operations in a basic well-defined manner. Making up the rules as you go is very difficult on personnel, and the rules will be difficult to learn, apply, and enforce. Failure to provide training can be a true liability for an organization. There are several things that are guaranteed for an organization that does not maintain a proactive and aggressive program. The list includes:

Injured personnel/fatalities
Accidents/damaged equipment
Poor customer service
Noncompliance with regulations or standards
Monetary loss due to excessive premiums

One of the most important training topics with which all personnel must be properly and thoroughly familiar is the organization's incident management system. An incident management system is defined as method for managing emer-

Training of Personnel

VIRGINIA BEACH FIRE DEPARTMENT	SOP ST 1
Training and Certification	09/01/93

VIRGINIA BEACH FIRE DEPARTMENT CERTIFICATION MATRIX

DEPUTY & DISTRICT CHIEFS
Driver's License
Defensive Driving
Emergency Vehicle Operators Course (EVOC)
Cardiopulmonary Resuscitation
Fire Officer III
HAZ MAT Operations and Incident Command
Executive Fire Officer (N.F.A.)
Fire Instructor II (required for Officer III)
TQM Supervisor Training*

BATTALION CHIEF
Driver's License
Defensive Driving
Emergency Vehicle Operator's Course (EVOC)
Cardiopulmonary Resuscitation (CPR)
Fire Officer Level II
HAZ MAT Operations and Incident Command
Fire Instructor I
TQM Supervisor Training*

Battalion Chief/Training
   Instructor IV

Battalion Chief/Inspections
   Inspector III
   Investigator III

Battalion Chief/Special Operations
   HAZ MAT Specialist/Technician
   Technical Rescue Modules
   PIO Training
   EMT

COMPANY OFFICER
Driver's License
Defensive Driving
Emergency Vehicle Operator's Course (EVOC)
Cardiopulmonary Resuscitation (CPR)
Emergency Medical Technician (EMT)
HAZ MAT Operations
Fire Officer I
Fire Instructor I

Figure 11.4. A fire department training matrix defining qualifications and recertifications required for each position.

VIRGINIA BEACH FIRE DEPARTMENT	SOP ST 1
Training and Certification	09/01/93

Basic Supervisor Training*
Grievance Procedure Training*
Performance Appraisal Training*
TQM Awareness Training*
All Career Firefighter Requirements

Company Officer/Training
   Instructor III

Company Officer/Inspections
   Inspector II

Company Officer/Investigations
   Investigator II
   Automatic Weapons Training**

Company Officer-Safety
   Instructor II

Company Officer/HAZ MAT Team
   HAZ MAT-Specialist/Technician

Company Officer/Technical Rescue Team
   Technical Rescue Modules

MASTER FIREFIGHTER
Driver's License
Defensive Driving Course
Emergency Vehicle Operator's Course (EVOC)
Cardiopulmonary Resuscitation (CPR)
Emergency Medical Technician (EMT)
Firefighter Levels II
Driver/Pump Operator
HAZ MAT Operations
Rope Rescue Level I
Confined Space Awareness
Trench Rescue Awareness
Vehicle Rescue Level II

Master Firefighter/Training
   Instructor II

Masdter Firefighter/Inspections
   Inspector II

Master Firefighter/Investigations
   Investigator II
   Automatic Weapons Training

Master Firefighter/HAZ MAT Team
   HAZ MAT-Technician/Specialist

Figure 11.4. (continued)

VIRGINIA BEACH FIRE DEPARTMENT	SOP ST 1
Training and Certification	09/01/93

Master Firefighter-Technical Rescue Team
   Technical Rescue Modules

CAREER FIREFIGHTER
Driver's License
Defensive Driving Course
Emergency Vehicle Operator's Course (EVOC)
Cardiopulmonary Resuscitation (CPR)
Emergency Medical Technician (EMT)
Firefighter Levels II
Driver/Pump Operator
HAZ MAT Operations
Rope Rescue Level I
Confined Space Awareness
Trench Rescue Awareness
Vehicle Rescue Level II
S.O.P. Proficiency Test
Career Firefighter–Training
   Instructor II

Career Firefighter/Inspections
   Instructor II

Career Firefighter/Investigations
   Investigator II
   Automatic Weapons Training**

Career Firefighter/HAZ MAT Team
   HAZ MAT–Technician/Specialist

Career Firefighter/Technical Rescue Team
   Technical Rescue Modules

RECRUIT FIREFIGHTER
TRFA
Driver's License
Defensive Driving Course
Emergency Vehicle Operator's Course (EVOC)
Cardiopulmonary Resuscitation (CPR)
Emergency Medical Technician (EMT)
Firefighter Levels I–II
HAZ MAT Operations
Rope Rescue Level I
Confined Space Awareness
Trench Rescue Awareness
Vehicle Rescue Level II
Infection Control
New Employee Orientation*

Figure 11.4. (continued)

COMPREHENSIVE RISK MANAGEMENT PLAN

VIRGINIA BEACH FIRE DEPARTMENT	SOP ST 1
Training and Certification	09/01/93

PUBLIC EDUCATION SPECIALIST
Driver's License
Defensive Driving Course (DDC)
Cardiopulmonary Resuscitation (CPR)

SPECIALITIES

LADDER COMPANY
Rope Rescue II
Vehicle Rescue II
SABA Training
Aerial Operator's Course
Confined Space Rescue
Trench Rescue

TECHNICAL RESCUE
Rope Rescue III
Confined Space Rescue
Trench Rescue
Vehicle Rescue II
Helicopter Operations
SABA Training

Additional: (Not Required)
  Rescue Specialist
  Helicopter Rig Master
  Basic Building Collapse

HAZ MAT
HAZ MAT Awareness
HAZ MAT Operations
HAZ MAT Technician/Specialist
Radiological Response Training

VOLUNTEER

BATTALION CHIEF
Driver's License
Defensive Driving
Emergency Vehicle Operator's Course (EVOC)
Cardiopulmonary Resuscitation (CPR)
Emergency Medical Technician (EMT)
Fire Officer Level II
HAZ MAT Operations
Fire Instructor I

Figure 11.4. (continued)

Training of Personnel

VIRGINIA BEACH FIRE DEPARTMENT	SOP ST 1
Training and Certification	09/01/93

SENIOR FIRE OFFICER (With Fireground Command)
Driver's License
Defensive Driving
Emergency Vehicle Operator's Course (EVOC)
Cardiopulmonary Resuscitation (CPR)
Emergency Medical Technician (EMT)
Fire Officer Level II
HAZ MAT Operations
Incident Command Training (N.F.A.)
Maintain all requirements of the Active Firefighter

DRIVER/OPERATOR
Driver's License
Defensive Driving
Emergency Vehicle Operator's Course (EVOC)
Cardiopulmonary Resuscitation (CPR)
Driver/Pump Operator's Course (EVOC)
Firefighter Level II
HAZ MAT Operations
Maintain all requirements of the Active Firefighter

EMERGENCY VEHICLE OPERATOR
Driver's License
Defensive Driving
Emergency Vehicle Operator's Course (EVOC)
Cardiopulmonary Resuscitation (CPR)
Firefighter Level I
HAZ MAT Operations
Maintain all requirements of the Active Firefighter

FIREFIGHTER
Driver's License
Defensive Driving
Emergency Vehicle Operator's Course (EVOC)
Cardiopulmonary Resuscitation (CPR)
Emergency Medical Technician (EMT)
Firefighter Level II
HAZ MAT Operations

*Denotes Department of Human Resources Training Programs
**Denotes Department of Police Training Programs

Figure 11.4. (continued)

RECERTIFICATION REQUIREMENTS

CERTIFICATION LEVEL	CERTIFICATION HOURS	RECERTIFICATION HOURS	EXPIRATION
Driver's License			5 Years
Defensive Driving	8 hours	4 hours	3 years
Cardiopulmonary Resuscitation	8 hours	4 hours	1 Year
Infection Control	4 hours	1 hour	Annually
HAZ MAT Team		16 hours	Quarterly
HAZ MAT Awareness	16 hours	8 hours/OSHA	Annually
HAZ MAT Operations	48 hours	8 hours/OSHA	Annually
HAZ MAT Technician	80 hours		
HAZ MAT Specialist	136 hours		Annually
HAZ MAT Incident Command	48 hours	8 hours/OSHA	Annually
Radiological Response Training	24 hours		Annually
Emergency Medical Technician	110 hours	30 hours/test	4 Years
Emergency Vehicle Operator's Course (EVOC)	16 hours	Test	5 Years
Firefighter Level I	75 hours	Test	
Firefighter Level I	45 hours	Test	5 Years
Driver/Pump Operator	64 hours	Test	5 Years
Aerial Operator's Course	48 hours	Test	5 Years
Fire Officer I	116 hours	Test	5 Years
Fire Officer II	48 hours	Test	5 Years
Fire Officer III	Pilot		
Instructor I	24 hours	Test	5 Years
Instructor II	24 hours	Test	5 Years
Instructor III	24 hours	Test	5 Years
Inspector II	120 hours	16 hours/test	2 Years***/ 5 Years
Inspector III		Test	2 Years***/ 5 Years
Investigator II	120 hours	40 hours/test	2 Years***/ 5 Years
Investigator III	240 hours	Test	2 Years***/ 5 Years
Rope Level I	16 hours	Not required	
Rope Level II	32 hours	Not required	
Rope Level III	24 hours	Not required	
Confined Space Rescue	16 hours	Not required	
Trench Rescue	16 hours	Not required	
Vehicle Rescue Level I	16 hours	Not required	
Vehicle Rescue Level II	16 hours	Not required	
SABA Training	8 hours	Not required	
Helicopter Operations	8 hours	8 hours	Biannually
Automatic Weapons Training	24 hours	8 hours	1 Year
Technical Team		16 hours	Quarterly
Rescue Specialist	20 hours		
Helicopter Rig Master	24 hours		
Basic Structure Collapse	20 hours		
Public Education Specialist	Test		5 Years

***Denoted Enforcement Powers

Figure 11.4. (continued)

gency operations, based upon organizational procedures and responsibilities. All personnel must understand their role and responsibilities in these procedures. Anyone by chance of positioning could find himself or herself initially in command of an incident. If that person is not trained on the use of the system, there could be more problems, especially in the area of safety. One of the most vital components of an incident management system is the accountability of personnel. Handling an emergency incident is difficult. If personnel turn into victims, that compounds the problem and catch-up is even more difficult. Personnel have to be trained on the use of the incident management system, especially personnel who function as incident commanders.

RISK MANAGEMENT AND TRAINING

In this chapter, we have attempted to identify valid reasons for providing training and education to personnel. All the reasons cited—from legislative mandates to humane considerations—are sound and legitimate. The fire service has witnessed a reduction in firefighter fatalities in the early 1990s. Occupational injuries have remained reasonably constant, based on statistical data collected by agencies such as NFPA and the IAFF. The intent here is that fatalities and injuries be reduced or eliminated through commitment from management, an aggressive safety and health program, compliance issues, and forecasting. All of these components are part of an overall risk management program.

Commitment from Management

An organization, whether it be a fire department, widget manufacturer, or little-league baseball team, must have the support of its management. There must be guidelines established by management that detail their philosophy and commitment to the training and education program. The intent of the organization is to train and educate personnel to the best of the organization's ability, based upon the safety and health of personnel, customer service delivery, and compliance with mandates. Management must convey this message to personnel in order for them to understand management's position. How many times have we heard personnel ask why they have to attend training, or remark: "We were taught this stuff in recruit school 10 years ago"? To the individual member, training and education may not seem important or necessary, but in truth they are.

Safety and Health Program

Safety has to be an integral part of any training program, and training and education must be integrated into the safety and health program. The organization has the responsibility to ensure that the training program incorporates safety into all facets of training for recruits, firefighters, and officers, and any other type of training provided. In the past, though it may not have been spelled out as such, safety has been included in training for personnel. Safety has become the mainstay of training programs and is highly visible throughout the process. This is true for recruit training, emergency medical technician certification, and fire officer training and certification.

In order for a safety and health program to succeed, personnel must be trained and educated to comply with the requirements. As with any policy or procedure, if the program is sent out without there being proper training, the success and/or compliance will be poor. Also, training on the safety and health program is not a one-time issue. The process should be revisited during in-service training, officer training, company drills, or any other avenue that is convenient for personnel.

Compliance Issues

Training may be instituted based upon legislative mandates that will describe necessary components of a program. These mandates will apply to issues such as: who has to be trained, how often training must be conducted, what are the training goals and objectives for the particular topic, and who is certified or qualified to conduct the training. Mandated training programs usually have a deadline for completion of training, so it is important to comply with this requirement.

Mandated training must be scheduled to integrate into the department's training timetable. Proper documentation of course completion must be maintained in the training files by the organization. In the event of a fatality or serious injury, this information will be requested as part of the investigation. The easier the information is to locate, the better.

Forecasting

Forecasting, from a risk management perspective, can be utilized in many ways. From the training standpoint, it can be used to "forecast" training requirements. Based upon recognized trends and the frequency and severity identified in a organization's annual accident and injury summary, specific training issues can be forecast. Information provided by the health and safety officer will allow personnel to forecast training needs. Near misses also serve as an indication that

Training of Personnel

Figure 11.5. Effective training ensures that personnel will equate knowledge, skill, and ability with safety (photo by Martin Grube).

a problem exists and that steps need to be taken to correct this problem. Personnel other than the health and safety officer, such as supervisors, incident commanders, and training staff, may be in the position to identify training problems and recognize the need to implement training to correct these problems or potential problems.

Forecasting is a valuable tool that must be utilized by the appropriate personnel in the department. If no steps are taken to correct the problem, personnel could be seriously hurt, and damage could occur to organizational assets. Hopefully, we learn from our mistakes rather than ignoring the fact that a mistake was made and therefore gaining nothing from it.

CONCLUSION

The training and education of personnel comprise a necessary and continuous process and are vital to the existence of an organization. Without a competent and effective training program, the organization will not be able to function and provide the necessary services to its customers, both internally and externally. With an effective program, procedures are modified, regulations are revised and updated, and technology improves. A proactive training and education program will incorporate these issues, which will allow members to provide professional and proficient services to customers.

PART *3* | *Emergency Incident Risk Management*

PROFILE

Chapter 12: Pre-Emergency Risk Management

MAJOR GOAL:

To understand the concept of pre-emergency risk management and identify the essential components of the pre-emergency risk management process

KEY POINTS:

- Conduct a pre-emergency inventory within your organization.
- Answer the following questions: Do you have a written risk management plan? If so, when was the last time you reviewed and revised it?
- Answer the following questions: Do you have a written safety and health policy? If so, when was the last time you reviewed and revised it?
- Understand that the health and safety officer and/or the occupational safety and health committee may conduct this inventory for the organization.
- Understand that the inventory should be conducted at least annually.
- Determine whether your risk management "toolbox" contains these components:
  ~ Standard Operating Procedures
  ~ Effective training
  ~ Protective clothing and equipment
  ~ Preventative maintenance programs for tools and apparatus
  ~ Personnel accountability system
  ~ Incident management system
- Understand that the pre-emergency risk management inventory will be a significant part of the organization's risk management plan.

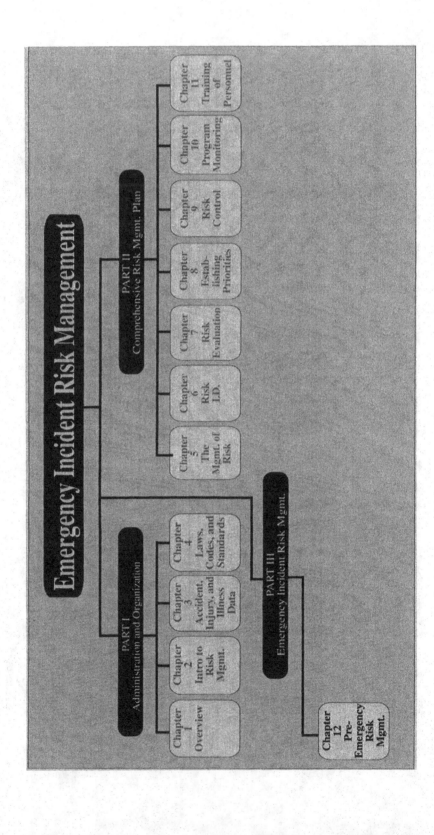

Chapter 12

Pre-Emergency Risk Management

INTRODUCTION

What Is Pre-Emergency Risk Management?

Pre-emergency risk management uses the classic risk management model approach presented in Part 2 of this book. Pre-emergency risks are those that are anticipated and that can be at least partially managed in advance of an emergency incident. This is the process that occurs prior to the response to emergencies, but that will make emergency scene risk management easier to perform. In earlier chapters, as we outlined the steps to effective risk management, we also addressed nonemergency and/or administrative risks.

The pre-emergency risk management elements must be identified and managed in order for an organization to safely and effectively conduct emergency operations. Therefore, pre-emergency risk management can be defined as a process that utilizes, prior to response, key safety and health elements that will reduce the hazards and risks involved during emergency operations and enhance customer service.

Necessary Components for Pre-Emergency Risk Management

Three initial components must be in place in order to establish a pre-emergency risk management plan. They are:

1. Written risk management plan
2. Written comprehensive safety and health policy
3. Health and safety officer function

These are not items that a department should have just for the sake of having a policy titled "safety" in the SOP manual. Each will affect the daily operations of the organization and impact it, based upon local factors such as philosophy, implementation, and management. Each of these pre-emergency risk management components will be expanded in detail so they can be clearly understood and recognized for their importance in the overall risk management process.

For many organizations, the component least likely to be already in place is the written risk management plan. Most organizations have used a written safety and health program for years and utilize a safety and health officer to oversee the safety and health process. It does not hurt, however, to re-evaluate each of these functions and amend them as needed so they can best serve the organization.

WRITTEN RISK MANAGEMENT PLAN

The process outlined in Part II of the book should, when completed, result in the outline of a written risk management plan. Risks are identified and evaluated, priorities determined, and control measures established. At this point, we shall utilize that process for handling pre-emergency risks.

A written risk management plan will be the result of a process that includes a review of the organization's policy and procedures. By formulating this plan, the organization is taking steps to avoid, control, or transfer risks, a process that will also help to protect against liability. The resulting plan will define how tasks, functions, or operations can be conducted in the safest manner possible. In the formation of this plan, the following functions must be considered from a risk analysis standpoint:

Administration
Facilities
Training
Vehicle operations
Personal protective equipment
Operations at emergency incidents
Operations at nonemergency incidents
Other, related activities

WRITTEN SAFETY AND HEALTH PROGRAM

The safety, health, and welfare of personnel is one of the most, if not *the* most, important responsibility of a fire chief or other top administrator, in the operation of an organization that responds to emergencies. The lack of those personnel who are out due to occupational injuries or illnesses disrupts the operation of the organization, and will have impacts that are both financial and moral.

The safety and health program outlines procedures for department personnel that, if followed, will enable them to perform their daily duties and responsibilities in a safe and effective manner. An effective written program places the responsibility and authority for safety on all personnel, and holds supervisors accountable for ensuring that personnel understand and comply with the requirements of this program.

There is no set method for developing a safety and health policy. The policy may be a single affirmation or may be part of a standard operating procedures (SOP) manual. Whatever method is used, all personnel must be provided with training and familiarization so they understand the intent and concept of this policy. Unfortunately, many policies written by an organization are placed in binders and put on a shelf, never to be seen again. If this happens to the safety and health policy, one should stand by for a lack of compliance by personnel, lack of support by supervisors, and general apathy.

NFPA 1500, *Standard on Fire Department Occupational Safety and Health Program,* Paragraph 2-3.1, describes that an organization shall implement written safety and health procedures that provide definitive measures for preventing and eliminating occupational accidents, injuries, illnesses, and fatalities. This policy requires that the organization's safety and health program meet the requirements of NFPA 1500.

An integral part of any safety and health program or effort is the statement of safety policy adopted by the leaders of the organization. A very simple, yet concise sample is included in Chapter 2. Regardless of the size of an organization, this kind of policy statement could work very well.

In addition, a statement of policy could address the responsibilities of the health and safety officer and the occupational safety and health committee. An example of this additional language may be as follows.

> The health and safety officer will serve as the program manager of the safety and health program. Under the direction of the Chief of the Department, the health and safety officer will be responsible for managing and maintaining the daily operations of the program.

Other statements might also include the occupational safety and health committee, and might state:

The occupational safety and health committee shall be established by order of the Fire Chief to serve as a component of the fire department's safety program. The intent of this committee is to provide assistance to the health and safety officer, conduct research and development projects on protective clothing, equipment, and apparatus, address department-wide safety problems, and serve as a voice of department members on safety issues. The membership of this committee shall be designed to include all interests within the department.

These statements are written so that any organization or department—career, volunteer, or part-paid—could adapt or utilize them as they are written.

As part of the monitoring of the risk management program outlined in Chapter 10, the safety and health program must be reviewed and updated on a periodic basis. The intent of this audit is to ensure that the safety and health program contains the necessary components and is meeting the stated goals and objectives of both the safety and health policy and the risk management program.

HEALTH AND SAFETY OFFICER FUNCTION

As stated earlier in this chapter, the safety, health, and welfare of personnel should be a primary concern for any organization. In order to provide an effective safety and health program for its members, the organization must develop, implement, and incorporate such a program in its daily management. To provide a consistently effective safety and health program, the program should be competently managed by an individual who has the qualifications and interest to do so. In a fire department, this would be the fire department safety officer, specifically the health and safety officer. In order to distinguish between functions of the safety officer and clarify roles, we shall call the administrative function or nonemergency safety officer the *health and safety officer*. The emergency safety officer—the one appointed to respond to emergencies—will be called the *incident safety officer*. This particular section deals with the functions of the health and safety officer. For details about the incident safety officer refer to Chapter 14.

What are the minimum qualifications for an individual to function as a health and safety officer? What are the duties and responsibilities of this individual as he or she functions daily in this position? National Fire Protection Association (NFPA) 1521, *Standard on Fire Department Safety Officer,* provides the basis for determining the minimum requirements and criteria for the positions of health and safety officer and incident safety officer. In addition to the requirements of NFPA 1521, each organization can also define its own criteria for this position.

The chief of the department ultimately carries the responsibility for the department's safety and health by incorporating a safety and health program and supporting a program manager. The health and safety officer is the program manager of the safety and health program and is responsible for the daily management of this program. As modifications to the program are needed, the health and safety officer must be the one responsible for ensuring that this is accomplished. An alternative health and safety officer must be appointed when the assigned health and safety officer is absent and be responsible for the duties that demand prompt attention.

The health and safety officer must meet the following criteria, identified in NFPA 1521, as well as any additional criteria developed by the department. Qualifications addressed in NFPA 1521 are:

Maintaining a knowledge of current federal, state, and local laws relating to firefighter safety and health

Maintaining a knowledge of potential safety hazards involved with structural firefighting and other emergency-scene operations, such as hazardous materials and confined-space entry operations

Maintaining a knowledge of health and physical fitness practices to improve the wellness of department members

Maintaining current knowledge of the principles and practices of safety management

Another factor to consider in the selection of a health and safety officer is the fire service experience of the individual. A thorough understanding of all fire department operations and activities is absolutely necessary. A person from a non-fire-service background will have to orientate himself or herself with this process in order to manage the safety and health program effectively. A department may require the individual to be an officer. As long as the individual has been given the authority by the fire chief to correct imminent hazards and conditions immediately, this will alleviate any concerns as to the authority and importance of the health and safety officer and the safety program. This authority must be implied in both emergency and nonemergency situations. At the incident scene, the incident safety officer must be able to alter, delay, or stop activities that present an imminent hazard. For nonthreatening situations, the incident safety officer can provide corrective actions through department policy and ensure that these problems are corrected.

The health and safety officer's enforcement of the department standard operating procedures is a critical issue. As these standards and regulations are developed and revised, the health and safety officer must be part of the revision process. It is crucial that these standards and regulations be revised annually and the health and safety officer be part of or at least review the changes to ensure that they do not conflict with mandated standards. As the national standards and regulations change, these changes must be incorporated into the department's SOP to ensure compliance. Laws, standards, and regulations that are mandated by requirements such as OSHA must be monitored and enforced by the health and safety officer.

The accident prevention program implemented through the occupational safety and health program and managed by the health and safety officer is an essential portion of the overall program. Direction must be provided through the written safety standard operating procedure as well as through verbal instruction. This can be accomplished through station drills, company in-service drills, and monthly training drills. The intent is to ensure that personnel are aware of safe work practices, both emergency and nonemergency. Many of the nonemergency work practices and procedures are governed under the OSHA regulations in 29 CFR 1910.

The health and safety officer position can be filled with a full-time or part-time person, depending on the needs of the organization. A large organization may need several individuals assigned to a safety division under the direction of a single health and safety officer. Other, smaller organizations may have a health and safety officer working a day shift as well as a health and safety officer assigned to each suppression shift.

Specific areas of responsibility may be assigned to individuals, based upon their expertise or knowledge. These areas may include:

Protective clothing and equipment
Research, development, and analysis
Worker's compensation
Risk management
Accident and injury reporting and recordkeeping
Accident investigation
Training and education
Apparatus and vehicle safety
Facility safety
Health maintenance
Infection control/EMS safety
Incident scene safety

The health and safety officer will typically be responsible for managing the day-to-day operations of the safety and health program. However, on an ongoing basis, this is not the responsibility of just one person. For the program to be successful, the health and safety officer must depend on the assistance of every supervisor and every member of the organization. An effective supervisor will realize that this program will make his or her job much easier in the long run, whereas poor supervisors who do not believe in the process will hurt not only themselves, but the personnel they supervise as well. Individual organization members must "buy into" this process, for it can help to protect them from needless injuries and accidents.

Pre-Emergency Risk Management 167

Figure 12.2. Effective training will reinforce the need to comply with departmental requirements for the use of personal protective clothing and equipment (photo by Martin Grube).

The health and safety officer will also require assistance from outside the organization, such as from an industrial hygienist for the evaluation of environmental hazards, or a medical professional for assistance with a communicable-disease exposure control program. Industrial hygienists have expertise that is essential to any safety and health program. They not only provide testing, but can also provide ideas on how reduce or eliminate some of these problems.

With the development and implementation of the health and safety officer function, the organization will reap many benefits. The benefits include a decrease in the number and severity of accidents, injuries, and illnesses, reduction in risk management costs or insurance premiums, and a decline in equipment and apparatus damage and replacement. This position requires commitment, dedication, knowledge, and initiative on the part of the individual(s) assigned to it. Because safety is good business, the health and safety officer function is a cost-effective one. The appointment and utilization of a health and safety officer is imperative to the safety and health of the members of the organization, especially when discussing pre-emergency risk management. The rewards are invaluable to the operation of the organization. General industry has shown us that the reduction in numbers and severity of injuries and accidents leads to a reduction in lost work time and workers' compensation costs, and is a direct result of an aggressive safety and health program. Fire departments utilizing this program are experiencing the same results as mentioned above. There is, hopefully, a growing trend within the fire service to develop, implement, and manage a safety and health program utilizing a health and safety officer. Remember, safety is good business.

TOOLBOX

As we continue to examine the pre-emergency risk management process, there are additional components that must be included in the process. In order to establish an effective risk management program, we shall need some tools to build this program. Put all of these tools together and you have the necessary components that fit into the control measures area of the risk management toolbox. Each component is a key element and, combined with the other components, produces an effective and safe department operation.

Components of the Toolbox

The components of the toolbox are:

    Standard operating procedures
    Effective training

Pre-Emergency Risk Management

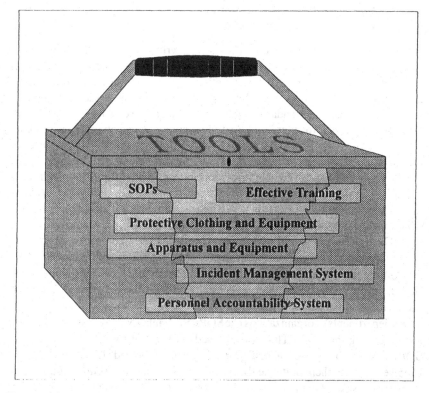

Figure 12.3. A toolbox with the necessary pre-emergency risk management tools.

Personal protective clothing and equipment
Apparatus and equipment
Incident management system
A personnel accountability system

As we discussed in presenting the classic risk management model, the risks that we encounter or potential risks that we may confront must be identified. We then evaluate these risks from the standpoint of frequency and severity or likelihood of occurrence, and then prioritize the risks based on past occurrences, accident and injury statistics, safety audit, and other components, and then we develop the control measures. Based upon these data, we can identify some vital components for the pre-emergency process that must be in place to make the program accountable, productive, and practical. Let's examine each of these components individually and discuss how they fit into the process.

Standard Operating Procedures (SOPs)

Standard operating procedures (SOPs) are written policies developed by an organization that define exact methods or activities performed by its members. These procedures affect only the operation of the organization that writes and adopts them. The requirements of these procedures must be based on recognized laws, regulations, and standards, which must meet or exceed the requirements. Standard operating procedures are the foundation on which an organization functions on a daily basis. The SOPs must cover all fire organizational operations, both emergency and nonemergency.

The basis of the standard operating procedures is quite simple, though individuals tend to make it more difficult at times than is necessary. As SOPs are developed and written, a training process must take place in order for personnel to understand what is expected of them. Once the SOPs become effective, they become enforceable. Once the training and education process has been completed, there is no excuse for noncompliance. It seems simple to say, in an SOP for interior structural firefighting, for example, that personnel shall wear full protective clothing and equipment including SCBA and facepiece. However, the fire service has a hard time with compliance.

As departments or organizations use SOPs, there must be a process for reviewing and amending the SOPs. This process needs to identify the effectiveness of the SOPs: Are they being used, and, if so, are they being followed? If they are not being followed, change them or delete them from the SOP manual. SOPs that are on the shelf for the sake of being on the shelf can come back to torment an organization. This is why it is very important to have a systematic process to review the SOPs on a regularly scheduled basis.

Effective Training

From a risk management prospective, we discussed in Chapter 11 the importance of having a training program. Without a training program and certification, it would be ridiculous to allow members to function at an emergency. From a pre-emergency risk management standpoint, this component is vital for ensuring consistency, efficiency, and safety. Without training, the fireground would be nothing more than an out-of-control mess.

The training process is an avenue for testing and evaluating new or revised standard operating procedures or policies. Training is also the approach for instituting and enforcing the safety process in a nonemergency mode or setting.

Personal Protective Clothing and Equipment

Prior to participating in any emergency operation, an organization must define the minimum level of protective clothing necessary to conduct business. This

Pre-Emergency Risk Management 171

Figure 12.4. The type of incident will dictate the proper type of protective clothing and equipment to use (photo by Martin Grube).

includes structural firefighting, vehicles, wildlands, or any other type of firefighting operation, as well as hazardous materials incidents, special operations, emergency medical services, or any other activity that requires the use of protective clothing and equipment. The organization has the obligation to ensure that the equipment provided is compliant and meets the intent for which it will be used. A key component is that personnel understand the use and limitations of the respective protective clothing. The garments and equipment are tested per certain criteria standards, which means that they have a limitation and are apt to fail once that limitation (for example, a defined temperature) is reached. The protective clothing has built-in safety factors, but they will provide protection for a very short period of time only. The maintenance and care of protective clothing is also important to the safety of the wearer. Poor maintainance of protective clothing and equipment leads to accidents and injuries. The manufacturer's recommendations should be followed with respect to cleaning and repairing the garments and equipment.

Apparatus and Equipment

The apparatus and equipment that are utilized for emergency operations must be properly equipped and properly maintained to maximize the safety of personnel. In order for this process to be productive, at least two components must be in place: a preventive maintenance program and a response SOP for "emergency driving" of apparatus.

The preventive maintenance program will ensure that routine maintenance and repairs are performed on apparatus on a scheduled basis. There will be criteria in place that allow apparatus to be placed out of service if certain conditions exist (e.g., poor brakes or steering). Work conducted on the apparatus must be performed by certified personnel trained by the manufacturer.

Incident Management System

The management of personnel at an emergency incident is the key to an effective, efficient, and safe operation. The risks are too great to allow the incident to be managed in an aimless and chaotic manner. Through standard operating procedures, an organization must provide a system that effectively manages an incident, using such elements as essential decision making, tactical design, plan survey and modification, and command and control. The incident management system needs to be flexible, yet solid enough to function in the following situations:

Fires
Hazardous materials incidents
Aircraft emergencies
High-rise incidents
Special operations: water, high angle, trench, and confined-space rescue
EMS incidents: mass casualty, multivictim
Any other emergency that requires the implementation of an incident management system

Personnel Accountability System

As an incident scene is managed, maintaining the accountability of personnel is a crucial function of the incident commander. Being able to account for the location of each member at an emergency incident is imperative in the event a problem develops that requires the tracking of all personnel on scene. In the past, through the incident management system, we have tried to control the accountabil-

Pre-Emergency Risk Management 173

Figure 12.5. The larger the incident becomes, the more important personnel accountability becomes (photo by Martin Grube).

ity of personnel, but freelancing still exists. A personnel accountability system will not be successful unless all personnel buy into the program. The best-written personnel accountability system will not be worth its weight unless personnel are trained in the process, use it, and are held accountable for nonuse or noncompliance. From a risk management perspective, the personnel accountability system is an excellent control measure. Chapter 15 will explain in detail the personnel accountability system.

Improving the Toolbox

The pre-emergency toolbox has many devices that will establish an effective program if properly developed and utilized. There may be other components that need to be added to the toolbox from time to time, based upon risk evaluation. An excellent example of this is the development of a highway/traffic safety policy. Due to the significant number of fatalities and injuries involving personnel at these scenes, it is imperative to develop a policy or procedure that addresses scene safety for responders. This policy gives an organization a control measure to implement that can easily reduce the risk to members who operate on highways under emergency conditions.

The key objectives for any operation at the scene of a highway incident are:

Preserving life
Preventing injury to emergency workers/responders
Protecting property and the environment
Restoring traffic flow

This policy gives the incident commander the basic guidelines for operating at a highway incident that can involve several agencies. This is in lieu of an out-of-control situation in which each agency is operating independently. The intent is to provide a means of reducing the hazards and risk to personnel by utilizing standard procedures for controlling traffic, operating safely on the highway, and managing other potential roadway occurrences. It is extremely important that all activities that block traffic lanes be concluded as quickly as possible and that the flow of traffic be allowed to resume as soon as possible. When the traffic flow is heavy, scene clearance time can mean a great saving in reducing traffic back-ups and the probability of secondary incidents. Restoring the roadway to normal or near-normal condition as soon as possible creates a safer environment. Issues that this policy should address are response of personnel, operation of apparatus, arrival and parking of apparatus and vehicles, protective clothing and equipment, use of warning lights, visibility of personnel and apparatus, and leaving the scene.

CONCLUSION

On a pre-emergency basis, the primary components necessary to an organization are a written risk management plan, a written safety and health policy, and implementation of the health and safety officer. Coupled with the pre-emergency toolbox, this provides a better means of controlling identified risks. Developing, implementing, and managing each of these components will outfit the organization with a viable means of providing an effective service to the community being served, in addition to addressing the safety, health, and welfare of the members of the organization.

An emergency services organization encounters a variety of incidents on a daily basis. Most of the incidents are handled in a rather predictable manner, and have a rather predictable outcome. The large-scale incidents that are encountered will thoroughly test the emergency response capabilities of the organization. The point is that we must always use our pre-emergency risk management tools to control the small incidents. Therefore, when the "big one" comes along, the outcome will be more predictable.

PROFILE

Chapter 13: Principles of Emergency Incident Risk Management

MAJOR GOAL:

To develop a standard strategy utilizing risk management priciples that will produce a desired outcome of an emergency incident

KEY POINTS:

- Understand that a desired outcome of every incident is to mitigate the emergency.

- Understand that a desired outcome is to ensure that all personnel leave the incident in the same condition in which they arrived.

- Determine whether your organization has procedures for proper size-up of an incident.

- Determine whether your organization conducts pre-incident planning.

- Determine whether pre-incident plans are available and used at emergency incidents.

- Determine whether your department utilizes a tactical worksheet.

- Determine whether strategy allows for matching tactics to conditions.

- Learn not to overlook the value of judgment, experience, and common sense.

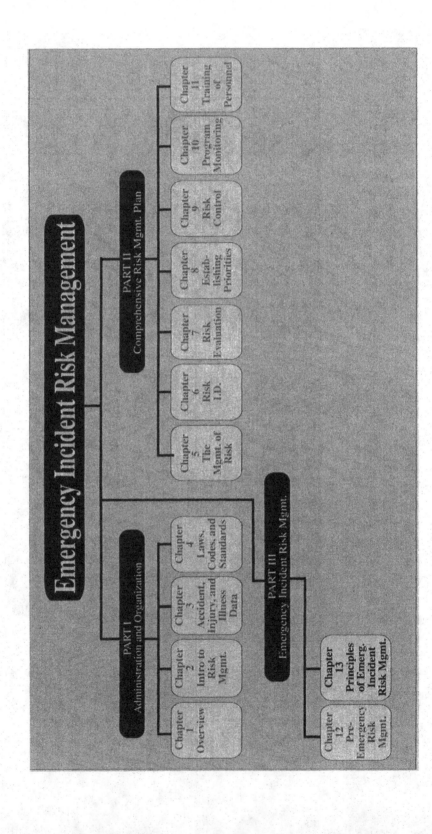

Chapter *13*

Principles of Emergency Incident Risk Management

EVALUATION OF CONDITIONS

In order to provide for the safety and welfare of personnel operating at an incident scene, the incident commander must systematically develop an incident action plan. The first step in this plan is to evaluate the conditions of the incident based upon:

- The type of incident
- Weather conditions
- Time of day
- Type of occupancy
- Life hazard
- Number of units responding
- Any additional information available from dispatch

This pertinent information will provide personnel with an initial evaluation of the incident to be coupled with reports from first arriving units and visual observation upon arrival. If possible, the incident commander should access the incident scene in a manner that allows the best possible view of the incident. This observation will provide information on actual conditions, location of the structure, exposures, location and placement of apparatus, and operations of personnel on-scene. This is a critical stage of the emergency incident. How well the incident commander digests this information and converts it into an incident action plan will greately impact the outcome of the incident.

Based upon this "size-up," the incident commander should determine the risk that personnel will be allowed to take. The incident commander must take time to consider the level of risk to personnel and the benefits to be gained by the tactical operations being employed. During firefighting operations, the decisions will have to be made more quickly than during special operations incidents, when time is not so crucial. If the incident commander does not take time to consider all the factors of an incident scene, the consequences could be hazardous to personnel and to the outcome of the incident. The initial stages of an emergency incident can be greatly hindered for an incident commander because of inaccurate or incomplete information. The incident commander must be able to receive information that will be processed into a strategy and then into tactical objectives. As the incident commander receives feedback from personnel, the incident action plan may have to be revised to reflect the changing conditions of the incident.

PRE-EMERGENCY RISK MANAGEMENT

Pre-incident planning is an important aspect of the pre-emergency risk management process. Often overlooked, determined as busy work, or deemed unimportant, pre-incident planning is critical to the safety and health of fire service personnel. Regardless of the size and type of community, each department needs to utilize a pre-incident planning program. This program allows personnel to visit and survey occupancies that are classified by various levels of hazard. Personnel become familiar with building construction, on-site hazards, and built-in fire protection systems particular to these occupancies. Past scenarios have shown that all too often a fire department was unfamiliar with known hazards associated with particular buildings, thereby hindering the ability to control the fire and endangering firefighters.

PRE-INCIDENT PLANNING PROGRAM

An intergal part of the pre-emergency risk management process is pre-incident planning. Pre-incident planning is a means of preparing for an emergency incident, given the known hazards associated with particular occupancies within a jurisdiction. The primary function of the pre-incident plan is to assist the incident commander with vital information that will be needed during the critical time of an incident. The information that is provided on the pre-incident plan has great bearing on the successful outcome of an incident and the safety of personnel operating at that incident.

Principles of Emergency Incident Risk Management 181

Figure 13.2. Firefighters can gain valuable information about a building during a pre-fire plan assessment (photo by Martin Grube).

Pre-incident plans need to be completed for those structures and occupancies that can be considered target hazards. A target hazard is defined as a site-specific feature such as life or occupancy, property value, product storage, conditions, fire load, and any other feature that could impede or hinder normal operations. Water supply problems, exposures, potential for delayed response, or any other special hazard related to the site characteristics of the structure or occupancy must be included as part of the target hazard criteria.

The pre-incident plan should be developed in a manner that assists the incident commander from the command post. From a pre-incident planning perspective, the following questions have to be answered:

Is this occupancy considered a target hazard?
What information is pertinent for the IC so that he/she can effectively contain this incident?
What unseen hazards could cause a potential problem in controlling this incident?
Are there any special resources that may be needed (e.g., technical rescue, hazardous materials)?

Pre-incident planning takes place in a nonemergency mode so that personnel can visit an occupancy and develop a pre-incident plan based on the information that the incident commander will need in the event of an emergency. The pre-incident plan process is usually conducted during the daylight hours, in a slow and methodical manner, and gives firefighters a chance to become familiar with an occupancy prior to a fire or emergency. This pre-emergency examination of the premises is imperative to the safety and health of firefighters during fireground operations.

Some departments routinely conduct pre-emergency drills at target hazards to familiarize and train personnel with respect to significant hazards or risks present at the facility or occupancy. Examples of target hazard facilities include hospitals, nursing homes, chemical plants, refineries, and any other type of facility or occupancy that has a potential for high risk of fire or life loss. As companies train, pre-incident plans should be a part of the training to ensure that the preplans are current and have correct information. The pre-incident plans should be maintained in such a manner that they are accessible during an emergency. Information that is stored in a notebook or binder that cannot be located during an emergency is useless to the incident commander, making the process a waste of time.

From a management point of view, the better pre-incident plans are managed, the better the process works. Usually, first-due companies maintain their pre-incident plans of their first-due area target hazards. It is impossible to expect a chief officer to maintain all of the pre-incident plans for a department. The plans can be filed alphabetically, by occupancy name or by address. Some departments are using mobile data terminals (MDT) that will store, access, and display pre-incident plan

Principles of Emergency Incident Risk Management   *183*

information quickly and easily. An important goal of this whole process is to make it practical to use and to make sure it is used. Once an emergency occurs, it is too late to gather all the information that will be needed to properly control an incident.

TARGET HAZARD CLASSIFICATION

Major target hazards should include:

Health care facilities
Lumber yards
Shopping malls
Correctional facilities
High-rise buildings (three or more floors)
Industrial facilities
Any other high-occupancy facility as determined by the battalion or district officer
Hazardous-materials sites (to be inspected by the hazardous materials team)

Other target hazards should include:

Large places of public assembly (500 or more)
Bulk storage sites for flammable or combustible solids, liquids, or gases
Schools
Shopping centers, regardless of size, that have at least one large "core" store, or for which there is an occupancy with an unusual hazard
Hotels and motels
Apartment or condominium complexes (there should be access or apparatus placement considerations, and floor and roof plans)
Churches (only if size or design dictates)
Any other high-occupancy facility as determined by the company or chief officer

These categories do not override the company or battalion officer's judgment as to the hazard of a particular occupancy as determined by the potential for significant life and/or fire loss.

COMPLETING THE PRE-INCIDENT PLAN

Pre-incident planning should be an annual process that is a company-level activity. Once a target hazard is identified and preplanned, a review of existing pre-incident plans should be conducted as well. The pre-incident plan should include:

- Data sheet
- Site plan
- Floor plan
- Roof plan

Data Sheet

This part of the pre-incident plan provides detailed information about the occupancy. Because there is a degree of complexity to most structures, information presented in the data sheet by personnel needs to be complete but should not be overwhelming. Pertinent information that is needed includes:

- Date and name of the shift/company completing the pre-incident plan
- Structure name and address
- Owner name and address
- Insurance information
- Emergency notification
- Fire protection equipment
- Type of utilities and their cutoffs
- Elevator information

Site Plan

The site plan provides a view of the occupancies and the surrounding area, detailing access to the structure, landscaping problems, and exposure problems. Personnel should utilize pre-incident plan symbols as much as possible in order to be consistent and make the information easy to read and understand. For large occupancies, such as schools, shopping centers/malls, and industrial facilities, the site plans may have to be divided into parts or groups, using natural or logical boundaries (e.g., major occupancy, fire wall, north end/wing). Criteria that will be needed include:

- Dimensions and distances (to be definitely noted)
- Compass direction (usually showing due north)
- Complete street names
- Fire walls
- Hydrant locations and size main
- Fire protection and location of hookups

Principles of Emergency Incident Risk Management 185

Figure 13.3. Information from a pre-fire plan can be invaluable during times such as this (photo by Martin Grube).

Drafting locations and accessibility to site
Bulk storage facilities
Hazardous-materials storage
Exposure problems
Power lines and other elevated obstructions
Landscape features particular to the site

Floor Plan

This part of the pre-incident plan shows the interior factors that can affect firefighting operations. The information should be as exact as possible. Again, the use of the pre-incident plan symbols is critical to detail features that are necessary or appropriate. Information that is pertinent to the floor plan includes:

Compass direction
Complete dimensions (shown and/or drawn to scale)
Room use
Exits and windows
Designation of front of the building
Special problems
Each floor with a different floor plan to be drawn on a separate sheet
All utilities: location of shut-offs and connections
Interior fire walls and fire doors
Elevators and elevator control rooms
Stairwells
Fire alarm systems and control panels
Fire suppression systems, control panels, and/or control rooms
Pre-incident plan symbols

Roof Plan

The roof plan should include information that is beneficial in the event roof operations are conducted. This could include rescue and ventilation plans. Information that is pertinent to the floor plan includes:

All features of the roof (to be noted and symbols utilized)
Dimensions, height of false fronts, parapets, mansards, and multilevel flat-roof
    structures
Compass direction

The pre-plan process must direct attention to characteristics that will affect tactical decisions and firefighting. Personnel have to understand and recognize the

Principles of Emergency Incident Risk Management   187

importance of conducting pre-incident planning. It is not a "make work" activity to keep personnel busy during the day. The information that is obtained during the pre-incident planning process is valuable to company personnel as well as the incident commander. From a pre-emergency risk management perspective, this is one of the necessary components that will affect the safety and health of personnel. As an organization develops its risk management plan, pre-incident planning is one of the initial components that must be utilized. During an emergency, the incident commander must ask for and utilize the pre-incident plan. A sample pre-incident planning SOP is shown in Figure 13-4.

PRE-INCIDENT PREPARATION

Besides the pre-incident planning process, there are several other vital components of a successful fireground operation. After a department has developed a pre-incident plan for a particular structure or situation, a process must be implemented that ensures that "first-due" companies tour these premises on a periodic basis. This allows companies to develop a tactical plan with regard to apparatus placement, accessibility, and general firefighting operations in the event of an emergency. Any changes or modifications made to the building can be noted at this time, ensuring that the pre-incident plan is updated. Fire protection features need to be noted and reviewed to ensure familiarity.

Tactical Worksheet

In order to manage an incident scene effectively, the incident commander must be afforded a means of tracking the location of personnel and their assignments, and recording pertinent information about the incident. The tactical worksheet provides the incident commander this means. It must be designed to fit the needs of the individual department. It must be easy to understand and be utilized by personnel during the course of an emergency incident. Some of the advantages of using a tactical worksheet are that it

- Records information in a regular format
- Serves as a reminder for the incident commander
- Provides standard form for all personnel to use
- Maintains useful information for the incident commander and staff

The tactical worksheet should include sections or areas for listing time of alarm, address, business name or type of occupancy, number and identification of first

VIRGINIA BEACH FIRE DEPARTMENT	SOP P 5
Pre-Incident Planning	10/15/94

PRE-INCIDENT PLANNING

PURPOSE

As a means to prepare for an emergency incident, the department has adopted a "pre-fire planning" program. The program goal is to identify alltarget hazards in the city and develop pre-fire plans for them. This program will provide structural information for and familiarization with identified hazards for our personnel.

SCOPE

This policy addresses the aspects of the Pre-incident Planning and Refamiliarization processes conducted by Suppression personnel.

HAZARD IDENTIFICATION

Pre-fire plans should be completed for those structures and occupancies that can be considered target hazards. A target hazard has specific characteristics such as life, property value, product (i.e. hazardous materials) or other characteristics which make it important that a pre-fire plan be prepared (generally speaking, day care centers, convenience stores, and fast food/restaurant establishments are not target hazards). Company Officers shall choose occupancies and structures that will meet the target hazard criteria. The Battalion Officer shall review the proposed list of properties to be pre-fire planned to insure they meet the criteria and are prioritized accordingly. A copy of the finalized list shall be sent to the District Chief's Office.

TARGET HAZARD CLASSIFICATIONS

MAJOR TARGET HAZARDS

- Health care facilities
- Lumber yards
- Shopping malls
- Correctional facilities
- Any other high occupancy facility as determined by the Battalion or District Officer
- Hazardous materials sites (to be identified and inspected by the Hazardous Materials Team)

TARGET HAZARDS

- Large places of public assembly (500 or more)
- Bulk storage of flammable/combustible solids, liquids, or gases
- Schools
- Shopping centers:
  - Any shopping center, regardless of size having at least one large "core" store
  - Any size shopping center if there is an occupant with an unusual hazard
- High-rise buildings—three (3) stories and up
- Hotels/Motels
- Churches (only if size or design dictates)
- Any other hazard occupancy as determined by the Company or Chief Officer
- Apartment/condominium complexes: only an expanded site plan is required unless there is a special need to include floor and roof plans.

Figure 13.4. A standard operating procedure for conducting a prefire planning assessment.

Principles of Emergency Incident Risk Management 189

VIRGINIA BEACH FIRE DEPARTMENT	SOP P 5
Pre-Incident Planning	10/15/94

These categories do not override the Company or Battalion Officers' judgment as to the hazard of a particular occupancy as determined by the potential for significant life and/or fire loss.

FREQUENCY
Pre-incident planning will be performed as an auxiliary company activity on an annual basis from July 1 to June 30. It is not the intent of this policy that structures and occupancies *not* identified as target hazards be pre-fire planned as a "make work" activity. If all identified TARGET HAZARDS have had pre-fire plan information completed, then a review of existing pre-fire plans should be conducted as part of the Refamiliarization process.

CONSIDERATION BEFORE PRE-FIRE PLANNING

Like any activity, pre-fire planning requires some planning and consideration before the task is attempted to insure success. The following represents various points that are associated with pre-fire planning which should be followed when engaged in this activity.

1. Permission to conduct pre-fire plans should be obtained in advance of the actual activity (usually a phone call to the owner or manager will suffice). Explain to the contact person the need for pre-fire planning and how it will help them in the event of an emergency at their facility.
2. Before pre-fire planning, personnel should be prepared to conduct a thorough and accurate pre-fire plan.
3. Class B uniforms shall be worn during all pre-fire planning activity.
4. Carry notebook, straight edge, and other materials so that notes and measurements can be kept.
5. The inspection team should consist of no more than three individuals. This is a rule-of-thumb and should be compared against the size and nature of the occupancy and the time required to conduct the pre-fire planning.
6. Personnel are reminded to be extremely courteous to occupants; if occupants are uncooperative, do not press for cooperation.
7. During pre-fire planning activities, apparatus is not to be left unmanned. For some companies this means that volunteers or personnel from other companies will have to help in pre-fire planning efforts.
8. The pre-fire plan is not an inspection. Problems identified should be handled through the Inspection Bureau.
9. Firefighters should be familiar with this SOP and the materials related to pre-fire planning.
10. Before leaving the site, check all information for completeness and accuracy to minimize repeat contacts to obtain missing information. Take along a data sheet to insure you get the information needed.

COMPLETING THE PRE-FIRE PLAN

GENERAL INFORMATION
The following instructions should be followed when completing all pre-fire plans:

Figure 13.4. (continued)

VIRGINIA BEACH FIRE DEPARTMENT	SOP P 5
Pre-Incident Planning	10/15/94

1. All required information should be provided.
2. Drawings should be neat with adequate dimensions. If your drawings are made to scale, include the scale utilized.
3. Those situations which are not covered in the SOP but in your opinion are important to the pre-fire plan should be noted on the plan.
4. Standard symbols have been established. Unusual situations should be noted so that new symbols can be developed when needed.
5. Compass directions should be noted on all diagrams.
6. Apartment complexes are no longer required to be completely pre-planned unless structural conditions dictate. An abbreviated single sheet apartment complex pre-plan form is available—side 1 includes all pertinent information about the structures, side 2 is a scale drawing site plan.

DATA SHEET (See Example)
This part of the pre-fire plan provides detailed information about a variety of facts dealing with the structure. Since structures vary in complexity, discretion should be used when completing this portion of the pre-fire plan. Reference materials related to building construction can be found in the station library.

SITE PLAN (See Example)
The site plan is an overview of the surrounding area (at least 100′ in all directions surrounding the structure being pre-planned). The accesses, landscaping problems, exposure problems, etc., are some of the factors that must be taken into consideration in this diagram. Standard symbols have been developed and should be used when possible. Deviations should be noted on the original drawings. For Shopping Centers/Malls, the site plan may be subdivided into major occupancies and/or groups of occupancies using natural or logical boundaries, i.e. fire walls, east wing, etc.
Criteria for completeness:
1. Building dimensions and distances should be clearly marked
2. Map will have compass directions
3. Streets names and locations are to be provided
4. Identify hydrant locations, main size, standpipe hookup, etc.
5. Identify drafting locations, distances and access problems
6. Show location of bulk storage areas and other hazardous materials
7. Exposure problems should be included
8. Show power lines and other elevated obstructions
9. Landscape features related to the site should be included when appropriate
10. Fire walls

FLOOR PLAN (See Example)
This part of the pre-fire plan should illustrate the interior factors which may affect fire operations. The diagram should be as accurate as possible. Standard symbols are to be utilized where possible; other features deemed necessary and/or important may be included.

Exception: Shopping Centers and Malls, the floor plan of each individual company need not be shown. Large and/or anchor stores should be shown individually with appropriate floor plan data noted; however, smaller occupancies may be shown in groups, i.e. fire wall-to-fire wall indicating basic building features.

Figure 13.4. (continued)

VIRGINIA BEACH FIRE DEPARTMENT	SOP P 5
Pre-Incident Planning	10/15/94

Criteria for completeness:
1. Compass direction
2. Complete dimensions shown and/or drawn to scale
3. Room use
4. Exits and windows
5. Designate front of the building
6. Special problems/irregularities
7. Each floor with a different floor plan should be drawn on a separate sheet of paper
8. All utilities-shutoffs and connections
9. Interior fire walls and fire doors
10. Elevators and elevator control rooms
11. Stairwells
12. Fire alarm systems and control panel(s)
13. Fire suppression systems and control panels and/or control room(s)
14. Use symbols wherever possible

ROOF PLAN (See Example)
The roof diagram should provide information relevant to operations that might occur on the roof of a occupancy, including ventilation and rescue.
Exception: (See floor plan)
Criteria for completeness:
1. All features on the roof should be noted and symbols utilized when possible.
2. Dimensions should be included—especially the height of false fronts, parapets, mansards, multi level flat roof structures.
3. Compass direction should be included.

PRE-FIRE PLAN PROCESS

A copy of the completed pre-fire plan should be made and retained by the Company (to be placed in the pre-fire plan book on the Engine) and a copy made for the Battalion Officers, with the original forwarded to the District Chief's Office. A pre-fire plan master list will be developed and updated annually, and should be utilized by Battalion and Company Officers to prioritize and manage the refamiliarization process.

REFAMILIARIZATION PROCESS

GENERAL INFORMATION
The purpose of this process is to refamiliarize shift personnel with target hazards and insure pre-plans are complete and up to date. The process is carried out on an annual basis from July 1 to June 30. Forms to log the on-site visits will be completed for that time period and will be sent to the District Chief's Office by July 15. All changes to the pre-fire plans throughout the year shall be forwarded to the District Chief's Office for inclusion in the master file maintained at this office.

Major Target Hazards: On site visits will be conducted by *all three shifts* during the year. Responsibility for updating the pre-plans should be equally divided between the three shifts.

Figure 13.4. (continued)

VIRGINIA BEACH FIRE DEPARTMENT	SOP P 5
Pre-Incident Planning	10/15/94

Target Hazards: Only *one* shift does the on-site visit during the year, up-dates the pre-plan, and notifies the other two shifts of any changes to the pre-plan. The list of Target Hazards should be divided between the three shifts in July to allow for better planning during the year and assignments should be rotated from year to year to assure all station personnel become familiar with all the target hazards.

To implement a change in a given target hazard list requires a request for an addition or deletion of a specified occupancy by the initiating company. A short narrative outlining the requests, signed off by the Company and Battalion Officers, and forwarded to the District Chief's Office for review, will begin the process. Written notification back to the initiating company will indicate approval or denial of the requests.

PRE-INCIDENT PLANNING FOR HAZARDOUS MATERIAL FACILITIES

PURPOSE
As a means to prepare for a hazardous material incident and to comply with SARA Title III, Emergency Planning Community Right-to-Know Act (EPCRA), the department has adopted a "pre-incident planning" program for facilities that manufacture, use and/or store hazardous materials. The program goal is to identify all hazardous material target hazards in the city and develop pre-incident plans for these facilities and to utilize this information in the **Virginia Beach Hazardous Materials Response Plan.**

SCOPE
All facilities identified as hazardous material target hazards should be pre-planned and/or a refamiliarization walk through should be completed by the Hazardous Materials Response Team.

HAZARD IDENTIFICATION
Target facilities will be identified by the Haz Mat Response Team. Facilities utilizing Extremely Hazardous Substances (EHSs) as defined under SARA Title III, EPCRA should be identified via the Tier II Reporting process.

FREQUENCY
These pre-plans should be performed on an annual basis between July 1 and June 30 of each year so the information can be included in the annual update of the **Virginia Beach Hazardous Materials Response Plan.**

Reviewed By	FIRE CHIEF

Figure 13.4. (continued)

Principles of Emergency Incident Risk Management   193

alarm and second alarm assignments, exposures, utilized sectors, emergency notifications, and space for a diagram of the incident scene. The tactical worksheet used by the Virginia Beach Fire Department is shown in Figure 13.5.

Risk Management

From the pre-incident preparation stage throughout the initial attack stage, there should be essentially no risk to personnel. Through the utilization of policy and procedures, training, and practice, the department can eliminate risks to personnel that respond to emergency incidents regardless of the situation. The risk management process begins once crews have arrived on scene and are beginning to mitigate the hazard or control the incident. By implementing this process, a department is making an enormous step toward ensuring the safety and welfare of personnel.

Phases of this process include the initial response, response from station or other point of dispatch, and on-scene arrival. To reduce the risks in the response phase, a department must ensure that personnel are completely dressed and "belted" before the apparatus moves, and must ensure safe driving of apparatus and vehicles under emergency response conditions and safe operations once on scene at the incident. Departments must address these issues in the department's risk management plan. Personnel must realize that if they respond to an incident and are themselves involved in an accident, they have not accomplished their objectives and did not provide the service their customers expect of them.

Civil Disturbance

Fire department personnel do not belong in situations involving violent acts or potentially violent acts. These are situations when one must rely on the presence of the law enforcement community. Fire department personnel should never be put at risk by entering such an environment. If an incident deteriorates, they must withdraw until the situation can be controlled. It is necessary to develop protocol prior to encountering these incidents to ensure that personnel are aware of their responsibility to protect themselves. If crews are dispatched to an incident that involves violence, they must stage well away from the scene and remain there until law enforcement personnel inform them it is safe to enter the area. Good liaison and communications with the appropriate law enforcement departments will ensure that this occurs.

EMERGENCY INCIDENT RISK MANAGEMENT

City of Virginia Beach
INCIDENT MANAGEMENT WORKSHEET

Address: _____

Occupancy: _____

Incident No. _____

Time of Alarm: _____

ACCOUNTABILITY
Elapsed Time

MODE: 10 20 30 40 50 60
Offensive   ☐ ☐ ☐ ☐ ☐ ☐
Transitional ☐ ☐ ☐ ☐ ☐ ☐
Defensive   ☐ ☐ ☐ ☐ ☐ ☐
PAR   ☐ ☐ ☐ ☐ ☐ ☐

I.C.: _____

	1A	2A	3A	4A		BACKFILL	
DC						1	
BC						2	
MS						3	
E						4	
E						5	
E						6	
L						7	
TA						8	
R						9	
R						10	
R						11	
SQ						12	
SQ						13	
HM						14	
TE						15	
S						16	
BR						17	
						18	
						19	
						20	
						23	

BENCHMARKS	NOTIFICATION	EMS	HAZ MAT	TECH RESCUE
☐ Primary Search	☐ City Manager/Mayor	☐ Triage	☐ Evacuation	☐ Extrication
☐ Secondary Search	☐ Fire Chief	☐ Treatment	☐ Research	☐ Search
☐ Exposures	☐ Deputy Chief	☐ Transportation	☐ Entry	☐ Rescue
☐ Confinement	☐ Police	☐ EMS Staging	☐ Decon	☐ Rig Master
☐ Fire Control	☐ EMS	☐ Med. Comm.	☐ Security	☐ Rappel Master
☐ Fire Out	☐ Public Utilities	☐ LZ		☐ Helo Ops
☐ Salvage	☐ Public Works	☐ Extrication		☐ Equip. Officer
☐ Loss Stopped	☐ Outside Agencies			
☐ Ventilation	☐ Inv / Insp / Fire Ed			
☐ Water Supply	☐ DES			
☐ Utilities G / E / W	☐ State			
☐ Rehab	☐ Federal			

Figure 13.5. A tactical worksheet used for managing incident operations.

CONCLUSION

Preparation for response to various types of emergency incidents in our various communities will enable a department to perform efficiently and effectively as well as maintain the safety and welfare of responding personnel. A department must determine the types of potential incidents to which it will respond, based upon past history, and plan accordingly. Failure to preplan for incidents greatly increases the risk to personnel and makes the situation difficult to manage for the incident commander.

Personnel must recognize the importance of this process and use it to their benefit. Valuable information can be obtained as to the placement of apparatus, methods of fire attack and extinguishment, types and location of fire protection systems, and accessibility. From a firefighter safety and health standpoint, nothing is more important than having information and data that will reduce the risks to personnel and enhance their safety.

A tactical worksheet will support the incident commander, ensuring that all pertinent information is recorded and documented. Managing an incident scene without the use of a tactical worksheet will result in inadequate information, poor scene control, and missed benchmarks. Planning is the key.

PROFILE

Chapter 14: Incident Safety Officer

MAJOR GOAL:

To develop, implement, and appropriately utilize the incident safety officer function at all incidents, and to ensure that the organization's incident management system recognizes and utilizes this role

KEY POINTS:

- Understand how to train and educate key personnel to function in this capacity.

- Understand that incident safety officers serve as the on-scene risk manager.

- Understand that the incident safety officer works within the confines of the organization's incident management system.

- Understand that the incident safety officer monitors compliance with all organizational safety operating procedures.

- Understand that the organization defines the implementation criteria for the incident safety officer role.

- Understand that the incident safety officer monitors for unsafe conditions, unsafe acts, and other incident hazards.

- Understand that forecasting is a valuable resource to be used by the incident safety officer to conduct risk assessment.

- Understand that the incident safety officer conducts risk assessment based upon the following "savability scale":
  ~ a lot
  ~ a little
  ~ none

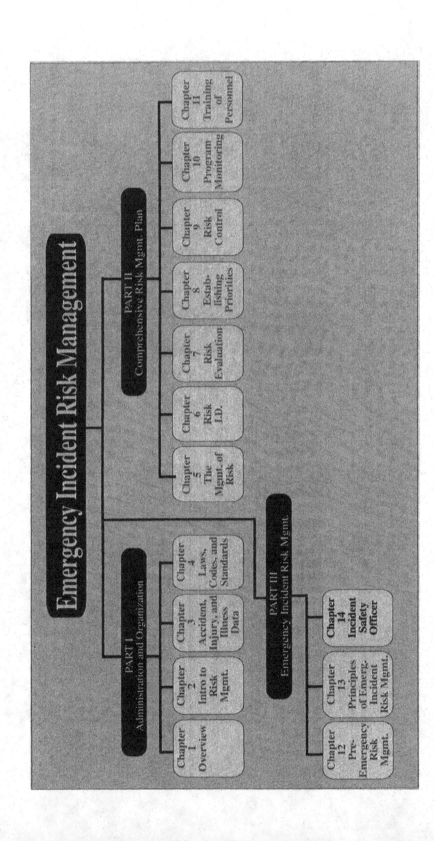

Chapter *14*

Incident Safety Officer

INTRODUCTION

The fire department safety officer is an integral part of all fire department operations during incident scene operations and during the daily nonemergency activities of the department. The purpose of that position is to focus an organization's efforts on the safety, health, and welfare of its members.

We can divide the safety officer position into two distinct functions, which are those of health and safety officer and incident safety officer. The health and safety officer is responsible for the administration and management functions of the safety and health program. The incident safety officer is responsible for the management of safety and health at an incident scene. The health and safety officer and the incident safety officer may be the same person or different people, depending upon organizational needs. The most important point is that the position is actively utilized and is recognized for its importance in completing an emergency operation safely and effectively. This chapter will focus primarily on the responsibilities of the incident safety officer.

It is imperative that functions of the incident safety officer be clearly defined and understood. This usually starts with the department's incident management system, which describes when and how the incident safety officer fits into the system. The "safety" sector can be refined to specifically describe the function of the incident safety officer at particular incidents. This can be done by a checklist or other means, as long as the goals are accomplished. We shall examine this in the next section of the chapter.

Responsibility and Authority

The safety, health, and welfare of personnel is the primary concern of any organization that provides an emergency service. In order to provide an effective safety and health program for its members, the department must develop and implement and incorporate such a program in its daily operations. To provide a continuously productive safety and health program, the program should be competently managed by individuals who have the necessary knowledge and skills.

The chief of the department or other administrator ultimately bears the responsibility for the department's safety and health. This individual must ensure that the safety and health program is developed, implemented, and proficiently managed in all situations. Though the incident commander has the overall responsibility for the safety of personnel at an incident scene, the responsibility is usually delegated to the incident safety officer.

Members selected to perform as incident safety officers must have a background that ensures that the right person is in the right job. As fire departments and EMS organizations utilize this position, they must ensure that members functioning in the position are knowledgeable about their responsibilities and that the duties are competently performed. Having an individual that is untrained or improperly trained could have serious consequences as the outcome of an incident. The member serving as incident safety officer must:

- Be a company, battalion, or chief officer
- Be familiar with the organization's incident management plan
- Have fireground command or incident management experience
- Be able to identify unsafe incident scene conditions and hazards
- Enforce procedures relating to protective clothing and equipment, and other regulations

It is imperative that the incident safety officer have respect and authority prior to initiation of this function. An officer is usually more qualified than a firefighter to perform the role of the incident safety officer because of his or her authority, experience, and knowledge. However, it is also important that the officers assigned to this function have received the proper training prior to fulfillment of this role. Having incident scene experience is a critical element for this position. The incident safety officer must have the knowledge and experience that comes of working within the confines of an incident management system. Allowing an inexperienced or untrained member to function in this capacity will significantly discredit the "safety" sector or, even worse, allow accidents and injuries to occur at the incident. The fact that the incident safety officer may be required to make instant decisions that could greatly affect the safety and health of personnel operating at the incident scene is justification enough for having qualified personnel in this position. Also, if personnel are not operating or following department policy, the incident safety officer must be cognizant that there is a problem and immediately correct it.

Annual accident and injury statistics reveal that more firefighters die and are injured at emergency scenes than any other location. Having an incident safety officer is one control measure for reducing the severity and frequency of accidents and injuries that occur at the incident scene.

RESPONSE CRITERIA

The functions of the incident safety officer must be a predetermined set of tasks as designated by the incident management system policy or the health and safety officer job description. Each organization will dictate the response policy of the safety officer to incidents. In some jurisdictions, the safety officer is a civilian and performs only administrative duties. Any response to an incident scene and operation of a "safety" sector will be made by a uniformed staff officer, such as an additional battalion or division officer. If the health and safety officer is a uniformed member and required to respond to emergency incidents, then policy would dictate response to an incident.

The following are conditions in which an incident safety officer would respond:

- Any second alarm or greater
- A working fire in a commercial structure
- Large-scale working incidents, such as wildland fires and hazardous-materials incidents
- Special operations, such as technical rescue incidents (trench rescue, high angle rescue, water, etc.) and mass casualty incidents
- Injuries to firefighters
- Accidents at an incident scene or incidents in which apparatus or equipment is damaged
- When the department responds mutual aid
- When requested by the incident commander

The incident safety officer must also be provided the latitude to respond to any incident. This does not mean that the incident safety officer responds to incidents just to drive up and down the street "blowing the siren." The response must be justifiable. If the incident safety officer, through monitoring the incident via radio, determines there are potential safety problems, then he/she can respond.

INCIDENT MANAGEMENT SYSTEM

Once there has been a response to an incident and the incident commander is on the scene, what happens next? The incident safety officer is part of the "command" staff and reports directly to the incident commander. The incident safety officer works solely for the incident commander as a consultant, counselor, or mentor who contributes a particular level of expertise to the operation. A competent incident commander will seldom operate without an incident safety officer close at hand. The incident commander is busy during the operations at an incident, developing strategies and managing tactical operations. Utilizing an incident safety officer allows a specialized look at the operations solely from a safety and health perspective.

The significance and magnitude of an incident may require the utilization of additional safety officers or assistant safety officers. These persons would be assigned to the safety sector under the direction of the incident safety officer. Examples of such incidents are:

high-rise incidents
mass-casualty incidents
hazardous-materials incidents

There may also be situations that require expertise from outside the fire department, in such areas as technical rescue scenarios or hazardous-materials operations. Additional safety officers may also be needed due to the specialized knowledge that may be required in order to mitigate the situation, enhance the safety of personnel, or provide an increased level of competency for the incident safety officer.

EMERGENCY AUTHORITY

In order for the incident safety officer concept to function properly, the incident safety officer must be granted authority to stop, alter, or suspend operations at an incident scene. NFPA 1521, *Standard for Fire Department Safety Officer,* provides the necessary verbiage. This allows the incident safety officer to "alter, suspend, or terminate" an operation if deemed necessary. This must be clearly spelled out in the department's incident management policy. The incident safety officer is an extension of the incident commander, serving as the eyes and ears of the incident commander during operations. Should the incident safety officer be faced with a situation that requires an immediate decision, he or she must take

Incident Safety Officer 203

Figure 14.2 An effective incident commander and command staff will ensure proper scene management (photo by Martin Grube).

immediate action to remedy the situation for the sake of the safety and health of personnel at the incident. The incident safety officer must communicate with the incident commander, advising him or her of this decision, because of effects on the remainder of operations. This activity is in no way a means of undermining the incident commander nor does it involve taking command of the incident, but is part of the responsibility given to the incident safety officer.

DUTIES AND FUNCTIONS

As we further define the duties and responsibilities of the incident safety officer, one resource is NFPA 1521, *Standard for Fire Department Safety Officer*. In Chapter 3, "Functions," requirements for incident scene safety are described:

> The incident safety officer responds to situations that involve high risk to personnel. The department must define the criteria for the response of the incident safety officer.
> At high-risk incidents, the incident safety officer must identify and eliminate safety hazards.

At the emergency scene, the incident safety officer shall work within the command structure and shall report to the incident commander.

The incident safety officer shall be granted emergency authority and responsibility to stop, alter, or suspend operations at an emergency scene.

The incident safety officer monitors operations at emergency incidents for compliance with department safety regulations. As necessary, the incident safety officer recommends corrective actions.

The incident safety officer must be involved in postincident critiques or analysis.

An incident safety officer must be able to identify incident scene safety hazards and conditions. Recognition of hazards requires immediate measures to correct these unsafe conditions and/or remove personnel from the area. When the incident safety officer arrives at the command post, an accountability "tag" must be given to the accountability officer, to be recorded in the accountability system. An incident safety officer does not have the authority to freelance at an incident, but there are times when the incident safety officer is separate from the command officer. For example, the incident safety officer may be on the opposite side of a building from the command officer, monitoring operations; "command" needs to know this. If the incident safety officer needs to go inside a structure, this individual needs to be with a company or another officer. It is imperative that the incident commander know where the incident safety officer is at all times. Upon arrival at an incident, the incident safety officer must meet "face to face" with the incident commander to determine what the immediate safety concerns are.

The key reason for meeting with the incident commander is to review the incident action plan. What are the strategies that the incident commander is utilizing to control the incident? What are the tactics being put in place to mitigate the incident? The incident safety officer must know what the incident commander's incident action plan is in order to formulate an operation plan for safety. This gives some direction to the incident safety officer.

INCIDENT SCENE MONITORING

Hopefully, the incident safety officer function will be utilized as dictated by the type of incident encountered. Let's examine some of the functions of the incident safety officer at a variety of incidents. There is absolutely nothing wrong with the incident safety officer's use of a checklist to monitor the incident scene. A note of caution, however, is to guard against the incident safety officer's becoming dependent on the checklist and forgetting to use common sense when "looking at the big picture." The incident safety officers should not be so absorbed in completing the checklist that the building is falling down around them or other incident safety problems exist in plain view without being recognized or addressed. The incident safety officers must utilize their knowledge, experience, and common sense, and have a positive attitude to ensure the safety, health, and welfare of personnel operating at an incident scene.

Fireground

The incident safety officer who functions on the fireground provides a level of expertise relating to protective clothing, equipment, and safety procedures. The safety sector must have the authority to migrate around the fireground to monitor and observe operations. This may include accepting and completing assignments for the incident commander, monitoring a specific area or situation, or looking at structural conditions. Examples of areas that an incident safety officer would monitor on the fireground include the following:

Use of full protective clothing and equipment by all members on the fireground
Structure condition and stability
Freelancing
Personnel accountability
Safety zones
Rest and rehabilitation of crew members

There may be other assignments or sectors that affect firefighter safety but they are not tasks performed by the incident safety officer. An example would be the "rehab" sector. The incident safety officer needs to ensure that "rehab" is operating with proper staff and facilities.

Some other examples of areas to monitor while on the fireground are:

Safety operation plan
Risk assessment:
   a lot
   a little
   none
Rapid-intervention crews
Crews ventilating a roof: whether there are:
   proper protective clothing and equipment
   proper tools
   a charged hose line
   at least two ways off the roof
Personnel operating directly over a fire

Personnel allowing a fire to get between them and their way out
Offensive and defensive operations going on simultaneously
Fire conditions (whether they have increased or intensified since the F.D. arrival)
Structural conditions (whether they indicate weakness or deterioration)
Establishing a safety perimeter (personnel that are supposed to be inside the perimeter must have proper protective clothing and equipment; everyone else must be outside)
Monitoring the condition of personnel, especially the first-alarm personnel

Emergency Medical Services (EMS) Incidents

Situations that may dictate the need for an incident safety officer might include:

A multi-vehicle accident
A mass casualty incident
A multi-patient accident
An industrial accident
Other situations, as determined by the incident commander

As with most medical emergencies, the primary focus is patient care, but as we have learned with firefighting operations, we must take care of our personnel as well. As the incident commander establishes an incident action plan, the safety of personnel must be included in these priorities. Our goal is to treat and transport the patients as quickly as possible, but the safety of personnel cannot be sacrificed in this situation. We do not want responders to become patients, thereby expanding and extending the incident. The incident safety officer may be appointed, in some scenarios, from a pool of on-scene personnel if an incident safety officer is not dispatched or at least until the incident safety officer arrives on scene.

The safety sector will serve, in this situation, to advise the incident commander of unsafe conditions, condition of personnel, and use of protective clothing and equipment. A checklist of areas to monitor are:

Establishment of a safety perimeter
Determination of the minimum level of protective clothing and equipment
Placement of protective hoselines in operation, based upon situation
Monitoring of scene conditions:
   stability of overturned vehicles
   leaking of fuel or other flammable liquids
   securing of possible ignition sources

lighting, especially at night
Safety of working environment and appropriate level of protective clothing and equipment
Infection control
Monitoring of conditions of personnel, both mental and physical
Presence of rapid intervention crews
Determination of whether one incident safety officer is adequate, based upon the size (area) of the incident or the magnitude (number of patients)
For long-term operations (three or more hours), the making of arrangements for extended "rehab" operations
Personnel accountability system in operation
Freelancing

Hazardous-Materials Incident

A hazardous-materials incident will require an incident safety officer and a hazardous-materials (haz mat) team safety officer. The haz mat safety officer may report through the incident safety officer or may report to the operations officer. The incident safety officer will be responsible for overall scene safety and the haz mat team safety officer will be specifically responsible for the haz mat team members and their safety.

When either of these functions are being filled, the safety checklist might look something like this:

Are personnel operating within the scope of their haz mat training?
Has a safety perimeter been established?
Are there adequate resources (personnel, protective clothing and equipment, and tools) available to mitigate the hazard?
As the incident safety officer or haz mat officer, are you confident in your abilities to manage the "safety sector"? If not, request assistance.
Are there resources available to manage the incident should it be long term (three hours +)?
Complacency of personnel for extended operations
Rapid intervention crews in place
Personnel accountability
Site safety operation plan
Zones are established (hot/warm/cold)
Decon area is established
Pre-entry medical
Product research conducted/delivered/reviewed
Site action plan conducted/reviewed

Leak/spill stopped
Decontamination and clean-up

Technical Rescue

Technical rescue incidents can include confined-space rescue, structural collapse, rope rescue (high-angle rescue), trench rescue, and water rescue. It is important to ensure that the "safety sector" is established and in place for these various incidents. The incident safety officer is not expected to have a complete understanding of all operations, but, in the event of technical rescue incidents, he or she may rely on the expertise of a team member, a technical expert such as an OSHA compliance officer, or someone who has a working knowledge of these operations. Such individuals can be assigned to assist the incident safety officer to ensure a safe operation and environment.

For technical rescue operations, the safety checklist will include:

Making the general area safe
Risk assessment
    a lot
    a little
    none
Rescue/recovery mode
Making the rescue area safe:
    Structural stability
    Atmospheric monitoring
    Ventilation
Enough personnel on-scene to safely conduct an operation
Site safety operation plan
Rapid-intervention crews
Proper personal protective equipment
Air supply, not >300 feet:
    Self-contained breathing apparatus (SCBA)
    Supplied-air breathing apparatus (SABA)
Explosion-proof lighting and communications
Decontamination
Personnel accountability
Monitoring of the mental and physical well-being of personnel
Complacency of personnel
Hazards to rescuers:
    Making trench lip safe
    Securing of utilities
Creating a safe zone

FORECASTING

Statistics show that the emergency scene generates more injuries and fatalities than any other location or function. Due to the fact that conditions change or deteriorate rapidly, the incident safety officer must continuously monitor the actions of personnel. The incident safety officer is concerned with the hazards or risks that create immediate threat to personnel or that may create a situation that becomes hazardous to personnel. A tool that the incident safety officer can utilize is forecasting of the conditions at the emergency as they relate to safety.

Unfortunately, the incident safety officer does not have a crystal ball to determine the outcome of an emergency incident with 100% accuracy. What the incident safety officer does have is the use of experience, training, safety prompts, knowledge, and instinct to stay a few steps ahead of the emergency and be able to anticipate developments that will impact the safety of on-scene personnel. Based upon the situation encountered, the incident safety officer has forecasting methods to foresee the future at an emergency.

Structural Fire Forecasting

The following is a list of considerations for the incident safety officer while conducting structural fire forecasting:

Building features and construction:
 Means of access
 Utilities (e.g., natural gas, electricity)
 Bowstring truss roof
 Indications of reinforcements
 Signs (visible or audible) of structural failure
 Signs of crew fatigue from locating and extinguishing hidden fire
Roof operations:
 Use of full protective clothing and equipment
 Presence of a charged hoseline
 Access to two ways off the roof
 Determining whether the fire has vented (if so, there is no reason to ventilate the roof)
 Testing of the roof before firefighters walk on it
 Ensuring that there are no "roof shepherds," which means that once the vent hole is cut, it is time to leave
Fire protection systems:
 Determining whether the sprinkler system is operating
 Consideration of the weight of the water from the sprinkler system

Knowledge of the number of past responses to this building for system problems or malfunctions

Determination of the presence of special extinguishing systems (e.g., carbon dioxide, halon) that will impact personnel safety

Accessibility for personnel:

Recognition of greater labor intensity in large buildings

Need for personnel to "chase" the fire due to difficulty in locating it

Age and condition of building:

Knowledge that older buildings usually do not have lightweight truss construction

Recognition that access and egress may be difficult

Possibility of early collapse due to new or pre-fab construction

Amount of fire involvement, noting that:

A big fire means defensive operations and less chance of survival for occupants. What are the risks to firefighters trying to make an interior attack?

Heavy fire involvement can mean early/sudden collapse or structural failure.

Safety, noting that:

The incident safety officer must consider that time the fire has been burning prior to his or her arrival.

Time works against us, not for us.

The adequacy of the water supply will impact the operations that are being undertaken.

Extreme heat will require early rehab and timed work rotations.

Cold weather creates additional problems for personnel, such as hypothermia, slips and falls, and frozen equipment.

Emergency Medical Operations Forecasting

The following is a list of considerations that the incident safety officer must evaluate with respect to EMS forecasting:

Communicable disease control:
  Hands
  Face and eyes
  Body
  Feet
  Respiratory protection
Violence:
  Against personnel
  Civil disturbance

Highway/traffic safety:
Interstate operations
Understanding that apparatus can be replaced but personnel cannot
Sufficient staffing
Police involvement
Incident management system

Special Operations Forecasting

Important features for Special Operations forecasting:

Noting that incidents in this category will last longer than other operations
Utilization of technical experts/consultants for the safety sector
Proper protective clothing and equipment
Rapid-intervention crews
Knowledge of operations that are being conducted

POST-INCIDENT ANALYSIS

The incident safety officer must be included during the post-incident analysis or critique. This is especially true for incidents that involve a fatality, serious injury, or an accident involving department apparatus or equipment. If none of the above occurred, the incident safety officer must still be able to discuss any safety concerns that occurred during the incident and how they can be handled differently at the next incident. This information should be available not only to the personnel at the particular incident, but to all department members. This process should be part of the safety training and education process, for which the health and safety officer is responsible. The training and education should also include information on accidents and injury investigations. The safety officer can provide information through drills, company inservice training, videos, and/or a monthly or quarterly newsletter.

CONCLUSION

The incident safety oficer must be given the authority to stop any operation that affects the safety and welfare of members operating at the incident. The incident commander is responsible for the safety of everyone at the incident, but it is the safety officer's responsibility to ensure that safety is in fact a paramount concern. The incident safety officer must monitor crew members, especially during extreme weather conditions. The members' welfare is critical to the success of the operation.

The correlation of the roles of the health and safety officer and the incident safety officer is a necessary one. The occupational safety and health program has to be managed on a daily basis, and this is a function of the health and safety officer. These duties may or may not involve responding to emergency incidents. If the health and safety officer does not respond, the organization must ensure that this function is performed by a shift safety officer or another staff officer. If the functions are separate, procedures must be in place to address training problems, investigations of accidents and injuries, and any other type of follow-up that is needed or required.

The primary goal of the safety and health program is to reduce the number and severity of accidents, injuries, and occupational illnesses. The health and safety officer and the incident safety officer contribute to this process in a variety of ways. Each position compliments the other, with the common goal of providing an effective occupational safety and health program.

PROFILE

Chapter 15: Personnel Accountability

MAJOR GOAL:

To develop, implement, and manage an accountability system that provides for the supervision and control of all personnel at an incident

KEY POINTS:

- Adopt a personnel accountability system that is compatible with the needs of the organization.

- Develop a written policy detailing the utilization and operation of a personnel accountability system.

- Ensure proper training and education of the accountability policy, to include a practical training application.

- Allow adequate time for acceptance of the personnel accountability system by members of the organization.

- Determine whether the accountability system is or can be integrated into the organization's incident management system.

- Understand that all command personnel must be held responsible for utilizing the system per organizational policy.

- Understand that all officers must maintain crew accountability at all incidents and be answerable to this regulation.

- Understand that all personnel must be responsible for utilizing the accountability system.

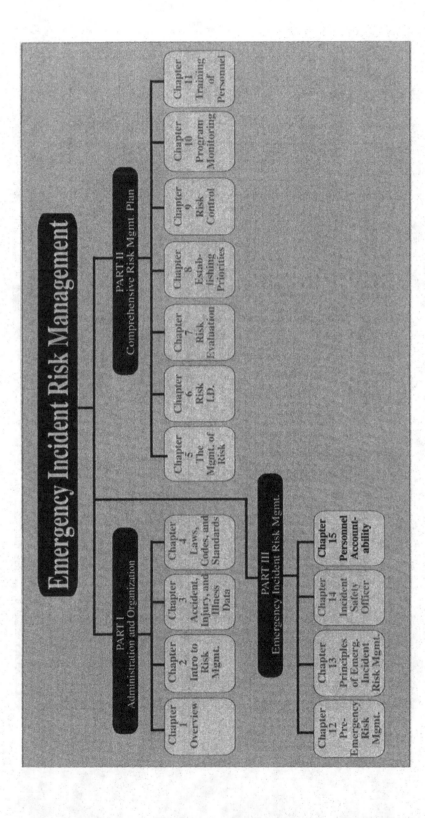

Chapter *15*

Personnel Accountability

INTRODUCTION

For fire service application, a personnel accountability program is defined as a tracking system that records the location and assignment of each member operating at an incident. Identification of the location of each member is documented by assignment on a vehicle or piece of apparatus (e.g., Engine 5) or by function (e.g., safety). Officers are accountable for the personnel assigned to their apparatus.

Personnel accountability is a proactive approach instituted for the benefit and welfare of personnel. NFPA 1500, *Standard on Fire Department Occupational Safety and Health Program*, Section 6-3, "Accountability," provides some basic requirements for utilizing an accountability system, which are as follows:

> Written procedures must be initiated for a personnel accountability system as defined in Section 4-3 of NFPA 1561, *Standard on Fire Department Incident Management System*, and that provides for the tracking and listing of all members that are operating at an emergency incident scene.
>
> The personnel accountability system shall incorporate local conditions and qualities during the establishment of the policy.
>
> All members operating at the emergency incident have the responsibility to participate and meet the requirements of the personnel accountability system.
>
> The incident commander is responsible for the management of the personnel accountability for the incident. The incident commander must enlist an accountability and inventory worksheet at the beginning of operations and maintain that system for the duration of the emergency incident.

The incident commander must maintain an awareness of the position and assignments of all companies and sectors.

Sector officers are responsible for directly supervising and accounting for all companies operating in their sector.

Company officers shall maintain constant awareness of the position and condition of all company members.

Each member is accountable for maintaining company integrity and remaining under the supervision of their assigned company officer.

Members are responsible for complying with the requirements of the personnel accountability system procedures.

The personnel accountability system is to be implemented and utilized at all emergency-scene incidents.

The fire department is required to develop and implement all system components to ensure that the personnel accountability system is effective.

The policy or procedures defining the personnel accountability system must provide for the use of extra accountability officers, based on the magnitude, complexity, or demands of an incident. These accountability officers work with the incident commander, sector officers, and company officers to maintain the tracking and accountability of members.

Figure 15.2. A riding list consisting of Velcro tabs ensures that the company officer maintains accountability for all personnel assigned to the apparatus (photo by Murrey Loflin).

Philosophy

From the standpoint of emergency incident risk management, the requirements listed above provide a solid basis for establishing a personnel accountability system. In order to develop a competent and sound incident action plan, all the ingredients must be available and utilized by the incident commander. As clearly stated throughout this process, the safety and health of personnel operating at an incident scene are the primary focus. In order for the incident management system to function at an emergency scene, there must be defined and written procedures that everyone knows and follows. The risks an incident commander will allow personnel to take or be exposed to are directly related to the hazards encountered at the incident scene. It is the responsibility of all personnel, including firefighters, company officers, sector officers, and the incident commander, to know and follow all procedures and policies adopted by the department. If a situation occurs, such as a partial collapse of a structure, a procedure must be in place that allows for quick and effective means of accounting for all members assigned within a sector or structure.

Past incidents show that the fire service has done an extremely poor job of tracking personnel at incident scenes, regardless of the type of emergency. Simply stated, an accountability system must be utilized at all incidents. If the system is used only for the large incidents, the system will become ineffective due to the lack of familiarity. For situations that require one or more companies to respond, the first arriving officer will be accountable for all personnel. For example, an engine company and a medic unit might both respond to a vehicle accident. If the situation subsequently requires additional units, the formation of a more formal accountability system will be required. The accountability system must be designed to allow for needs at the incident. It becomes very difficult to play "catch up" once an incident increases in magnitude.

Even though the fire service is provided with some excellent tools to accomplish this task (e.g., PASS devices, numerous types of accountability systems, incident management systems) firefighters continue to lose their lives because they are separated from their crews, become lost, or are freelancing.

Reasons for a Personnel Accountability System

Have you ever sent to a customer a package that had to be there by the next morning? The shipper provided you a tracking number for the package and a telephone number to call to confirm that the package arrived on time. From the time you sent that package, you were able to track it from point A to point B and points in between.

EMERGENCY INCIDENT RISK MANAGEMENT

Figure 15.3. Use of a personnel accountability system at all incidents is imperative for the safety of personnel (photo by Martin Grube).

Airlines now incorporate a system that can track luggage that is incorrectly routed or not sent at all. Using a bar-code system, they can locate luggage more quickly than before. We are better able to track a package that travels several thousand miles or luggage that has been sent on a wrong flight, than to track the location of firefighters at an emergency incident.

Weaknesses of the Personnel Accountability System

There are identifiable weaknesses in the personnel accountability system if all the players do not perform their roles all the time. This is not something that firefighters use only when they feel like it or when the "big one" taps in. The reasons that a personnel accountability system is not used include the following.

The system has not been written and implemented yet.
The system is not used by all personnel at the incident scene.
Personnel forget to use the system.
The personnel accountability system is not integrated into the incident management system.
The fact that NFPA 1500 requires a personnel accountability system is not seen as a reason for a department to use one.

The incident commander is "too busy" to employ the personnel accountability system and track the location of personnel at an emergency incident.
The system is considered too much trouble to deal with; there are more important issues that need to be addressed at an incident scene.
The "buy in" process by the troops is not complete.

Positive Aspects of the Personnel Accountability System

It is also significant to identify the postive aspects of using a personnel accountability system. The primary reasons are:

The system can geographically track and locate all personnel operating in the hazard zone.
Doing so reduces the likelihood that personnel will freelance.
Doing so maintains crew accountability.
Doing so allows the incident commander to locate a missing firefighter faster.

From a safety and health standpoint, the focus must be on the positive reasons for utilizing a personnel accountability system, rather than on the negative ones. The staff must give the process a chance, rather than assuming that the system does not and will not work.

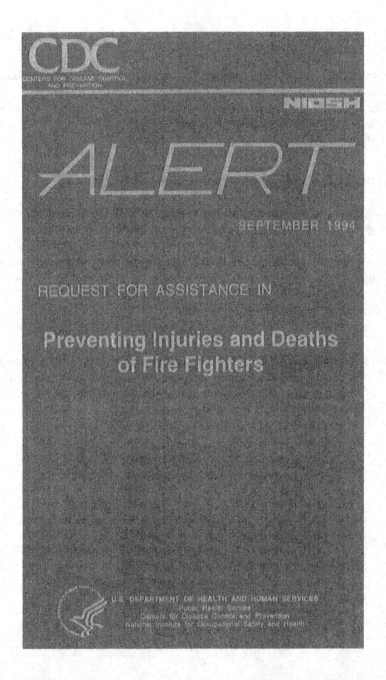

Figure 15.4. The National Institute for Occupational Safety and Health (NIOSH) published this Bulletin (reprinted here, pages 220-231, inclusive) to inform the fire service about the importance of safety programs and standard operating procedures during emergency operations.

DISCLAIMER

Mention of any company or product does not constitute endorsement by the National Institute for Occupational Safety and Health.

This document is in the public domain and may be freely copied or reprinted.

Copies of this and other NIOSH documents are available from

Publications Dissemination, DSDTT
National Institute for Occupational Safety and Health
4676 Columbia Parkway
Cincinnati, OH 45226

Fax number: (513) 533-8573

To order NIOSH publications or to receive information about occupational safety and health problems, call 1-800-35-NIOSH (1-800-356-4674)

DHHS (NIOSH) Publication No. 94-125

Figure 15.4. (continued)

NIOSH ALERT

 Request for Assistance in

Preventing Injuries and Deaths of Fire Fighters

> **WARNING!**
>
> Fire departments must review their safety programs and emergency operating procedures; failure to establish and follow these programs and procedures are resulting in injuries and deaths of fire fighters.

The National Institute for Occupational Safety and Health (NIOSH) requests assistance in preventing injuries and deaths of U.S. fire fighters. A recent NIOSH investigation identified four factors essential to protecting fire fighters from injury and death: (1) following established fire-fighting policies and procedures, (2) implementing an adequate respirator maintenance program, (3) establishing fire fighter accountability at the fire scene, and (4) using personal alert safety system (PASS) devices at the fire scene. Deficiencies in any of these factors can create a life-threatening situation for fire fighters.

NIOSH requests that the information in this Alert be brought to the attention of all U.S. fire fighters—including members of the largest metropolitan fire departments and the smallest rural volunteer fire departments—by the following: editors of trade journals and other related publications, safety and health officials, labor organizations, fire-fighting agencies, and insurance companies.

Photo by Glen E. Ellman (printed with permission)

Fire fighters wearing self-contained breathing apparatus (SCBAs) and other protective gear at the scene of a fire.

Figure 15.4. (continued)

BACKGROUND

Fatality data for U.S. fire fighters are collected by the NIOSH National Traumatic Occupational Fatalities (NTOF) Surveillance System, the International Association of Fire Fighters (IAFF), and the National Fire Protection Association (NFPA).

During the period 1980–89, 278 fire fighters died from work-related causes, according to data from the NTOF Surveillance System [NIOSH 1994b]. This figure includes only deaths from traumatic injury—not those from other causes such as heart attacks. The actual number of fire fighters who died is higher than reported by NTOF because methods for collecting and reporting these data tend to underestimate the total number of deaths [NIOSH 1993].

Data collected by the IAFF show that 1,369 professional fire fighters died in the line of duty during the period 1970–94 [IAFF 1994].

Data collected by the NFPA show that 280 fire fighters died and approximately 100,000 were injured in the line of duty during the period 1990–92 [Washburn et al. 1991, 1992, 1993].

CURRENT STANDARDS

OSHA and MSHA Regulations

State and local government employees are exempt from Federal OSHA standards. However, in the 25 States currently authorized by OSHA to run their own safety and health programs, all OSHA regulations apply to both public and private employees.

Current OSHA regulations that apply to fire fighters include 29 CFR* 1910.134 (Respiratory Protection) and 29 CFR 1910.156 (Fire Brigades). In 29 CFR 1910.134, employers are required to provide respirators suitable for the purpose intended and to establish and maintain a respiratory protection program. In 29 CFR 1910.156, requirements are listed for organizing, training, and equipping fire brigades established by the employer.

NIOSH and Mine Safety and Health Administration (MSHA) regulations [30 CFR 11] list the requirements for certifying respiratory protective devices, including self-contained breathing apparatus (SCBA).

Consensus Standards

The National Fire Protection Association (NFPA) recommends that all fire departments establish a policy of providing and operating with "the highest possible levels of safety and health for all members" [NFPA 1992]. Several NFPA standards apply to fire-fighting operations:

- NFPA 1404 specifies the minimum requirements for a fire service respiratory protection program [NFPA 1989].

- NFPA 1500 specifies (1) the minimum requirements for a fire department's occupational safety and health program, and (2) the safety procedures for members involved in rescue, fire suppression, and related activities [NFPA 1992].

- NFPA 1561 defines the essential elements of an incident management system [NFPA 1990].

---
*Code of Federal Regulations. See CFR in references.

Injuries and Deaths of Fire Fighters

Figure 15.4. (continued)

223

- Other relevant NFPA standards include NFPA 1971 (clothing), NFPA 1972 (helmets), NFPA 1973 (gloves), NFPA 1974 (footwear), NFPA 1981 (SCBA), and NFPA 1982 (PASS).

The American Society of Mechanical Engineers (ASME) has recommended the following standards for the control of elevators by fire fighters [ASME 1990a, b]:

- ASME A17.1—1990 Safety Code for Elevators and Escalators (as amended)
- ASME A17.3—1990 Safety Code for Existing Elevators and Escalators (as amended)

CASE REPORT: TWO DEATHS

On April 11, 1994, at 0205 hours, a possible fire was reported on the ninth floor of a high-rise apartment building [NIOSH 1994a]. This building had been the scene of numerous false alarms in the past. An engine company and a snorkel company were the first responders and arrived at the apartment building at 0208 hours. The engine company was the first on the scene and assumed command.

Five fire fighters from the two companies entered the building through the main lobby. They were aware that the annunciator board showed possible fires on the ninth and tenth floors. Lobby command radioed one fire fighter that smoke was showing from a ninth-floor window. All five fire fighters used the lobby elevator and proceeded to the ninth floor.

When the doors of the elevator opened on the ninth floor, the hall was filled with thick black smoke. Four of the fire fighters stepped off the elevator. The fifth fire fighter, who was carrying the hotel pack,[†] stayed on the elevator (which was not equipped with fire fighter control) and held the door open with his foot as he struggled to don his SCBA. His foot slipped off the elevator door, allowing the door to close and the elevator to return with him to the ground floor.

The remaining four fire fighters entered the small ninth-floor lobby directly in front of the elevator. One fire fighter stated that he was having difficulty with his SCBA and asked for the location of the stairwell. Another fire fighter said, "I've got him," and proceeded with him into the hallway, turning right. Later, one of the four fire fighters stated that he had heard air leaking from the SCBA of the fire fighter having difficulty and had heard him cough.

The remaining two fire fighters entered the hallway and turned left, reporting zero visibility because of thick black smoke. Excessive heat forced them to retreat after they had gone 15 to 20 feet. They proceeded back down the hall past the elevator lobby. There they encountered a male resident, who attacked one of the fire fighters, knocking him to the floor and forcibly removing his facepiece. The two fire fighters moved with the resident through the doorway of an apartment, where they were able to subdue him. One fire fighter broke a window to provide fresh air to calm the resident. At about the same time, the low-air alarm on his SCBA sounded. The other fire fighter was unable to close the apartment door because of excessive heat from the hallway. Both fire fighters and the resident had to be rescued from the ninth-floor apartment window by a ladder truck.

Fire fighters from a second engine company arrived on the scene at 0209 hours. They

---

[†]Two 100-foot lengths of hose. The hotel pack is also referred to as "standpipe pack" or "high-rise pack."

Injuries and Deaths of Fire Fighters

Figure 15.4. (continued)

observed a blown-out window on the ninth floor and proceeded up the west-end stairwell to the ninth floor carrying a hotel pack and extra SCBA cylinders. These fire fighters entered the ninth floor with a charged fire hose and crawled down the smoke-filled hall for approximately 60 feet (the hallway was 104 feet long) before extreme heat forced them to retreat. As they retreated, they crawled over something they thought was a piece of furniture. They did not remember encountering any furniture when they entered the hallway. In the dense smoke, neither fire fighter could see the exit door 6 feet away, and both became disoriented.

After the fire fighter from the first company rode the elevator to the ground floor lobby, he obtained a replacement SCBA and climbed the west-end stairs to the ninth floor. When he opened the ninth-floor exit door, he saw the two fire fighters from the second engine company in trouble. He pulled both into the stairwell.

When a rescue squad arrived at the scene at 0224 hours, lobby command could not tell them the location of the fire fighters from the first company. They proceeded up the west-end stairs to the ninth floor.

The rescue squad opened the ninth-floor exit door and spotted a downed fireman approximately 9 feet from the door. He was tangled in television cable wires that had fallen to the floor as a result of the extreme heat. The downed fireman was from the first engine company; his body may have been what the fire fighters from the second engine company encountered in the hallway. He was still wearing his SCBA, but he was unresponsive. The rescue squad carried him down the stairs to the eighth floor, where advanced life support was started immediately.

The rescue squad then entered the first apartment to the left of the exit door and found a second fire fighter from the first engine company kneeling into a corner and holding his mask to his face. He was unresponsive. The rescue squad carried the fire fighter down the stairs to the eighth floor where advanced life support was started.

Both fire fighters were removed within minutes and taken to a local hospital, where advanced life support was continued; but neither responded. Both victims died from smoke and carbon monoxide inhalation.

Both victims wore PASS devices; but because the devices were not activated, no alarm sounded when the fire fighters became motionless.

DISCUSSION

Many factors contributed to the deaths and injuries that occurred in this incident. The key factors were as follows:

- The first five fire fighters on the scene took an elevator to the floor of the fire—a violation of their department's written policy. Fire fighter entrapment in automatic elevators is a recognized hazard, and the elevators in this incident had no fire fighter control. ASME standards require fire fighter control for all elevators, and many elevator codes and installation practices were changed years ago to facilitate their safe use for fire fighting.

- At least one of the SCBAs leaked during this incident, and the respirator maintenance program appears to have been deficient. All four SCBAs tested by NIOSH failed at least two of five performance tests.

Injuries and Deaths of Fire Fighters

Figure 15.4. (continued)

- When the rescue squad inquired about the location of the first fire fighters at the scene, no one could account for them. Accountability for all fire fighters at the scene is one of the fire command's most important duties.

- PASS devices were worn but not activated by the two fire fighters who died. These devices should always be worn and activated when fire fighters are working at the fire scene.

CONCLUSIONS

Although many factors contributed to the deaths and injuries reported here, they might have been prevented if these essential steps had been taken:

- Following established fire-fighting policies and procedures
- Implementing an adequate respirator maintenance program
- Establishing fire fighter accountability at the fire scene
- Using PASS devices at the fire scene

These precautionary steps are well known to fire departments and fire fighters, but they require constant emphasis to assure the safety of fire fighters.

RECOMMENDATIONS FOR FIRE DEPARTMENTS

NIOSH recommends that fire departments take the following precautions to protect fire fighters from injury and death:

1. Establish and implement an incident management system with written standard operating procedures for all fire fighters. The system should provide for the following:

- A well-coordinated approach to the emergency
- Accountability of all fire fighters
- Overall safety of all fire fighters at the scene of the emergency

Train fire fighters in this system and provide periodic refresher courses to review policies and procedures. Fire fighters must always be fully aware of standard operating procedures and of their roles and responsibilities.

2. Develop and implement a written respirator maintenance program for all respiratory protective equipment used by fire fighters. Establish service and maintenance procedures and rigidly enforce them to provide respirators that are dependable and are constantly evaluated, tested, and maintained.

Include the following elements in the respiratory program:

- *Service checks.* Include daily, weekly, and monthly service checks in the standard operating procedures for servicing and testing SCBAs, cylinders, air quality, and air supply equipment. Such testing and servicing is extremely important in maintaining SCBAs for use in emergencies.

- *Approved respirators.* Use only respirators approved for use in hazardous atmospheres; maintain them in a NIOSH/MSHA-approved condition so that they are the equivalent of devices that have received a certificate of approval [30 CFR 11.2(a)].

Injuries and Deaths of Fire Fighters

Figure 15.4. (continued)

- *Training.* Train fire fighters in the use, care, and maintenance of respiratory equipment. Provide courses to review the fire department's policies and procedures for respiratory protection.

- *Recordkeeping.* Recordkeeping is a critical element of any respiratory protection program. Record the following items:

    — Results of the regular calibrations of the test equipment recommended by the manufacturer

    — Results of regularly conducted performance tests

    — Repairs made during routine preventive maintenance and necessary maintenance on SCBAs taken out of service.

    These records should include the SCBA and regulator identification numbers, test equipment identification numbers, dates of servicing, a description of the action taken (including parts replaced and part numbers involved), and identification of the repair person [29 CFR 1910.134; 49 CFR 173; NFPA 1989; NIOSH 1987].

- *Tracking system for SCBA cylinders.* Establish a tracking system for SCBA cylinders to ensure that they are hydrostatically retested and recertified (every 3 years for aluminum and composite cylinders, and every 5 years for steel cylinders) as required by the Department of Transportation (DOT) [49 CFR 179.34 (e)(13)] and NIOSH [30 CFR 11.80(a)].

3. Establish and implement a system of accountability that will enable the commander at the scene of the emergency to account for the location and function of each company or unit at the scene. Also use a standard personnel identification system that can rapidly account for each department member at the scene.

4. Employ a buddy system whenever fire fighters wear SCBAs. Fire fighters who wear breathing apparatus should never enter a hazardous area alone. Two fire fighters should work together and remain in contact with each other at all times. Two additional fire fighters should form a rescue team that is stationed outside the hazardous area. The rescue team should be trained and equipped to begin a rescue immediately if any of the fire fighters in the hazardous area require assistance. A dedicated rapid-response team may be required if more than a few fire fighters are in the hazardous area [Morris et al. 1994; NFPA 1990, 1992; NIOSH 1990].

5. Provide PASS devices and ensure that fire fighters wear and activate them when they are involved in fire fighting, rescue, or other hazardous duties [NFPA 1992].

6. Encourage municipalities to review and amend their elevator and life safety codes to require fire fighter control for all elevators with a total travel distance greater than 25 feet [ASME 1990a,b].

7. Guard against heat stress and other medical emergencies at the fire scene; provide cool water supplies, rest areas, and access to emergency medical personnel [NIOSH 1985, 1986].

RECOMMENDATIONS FOR FIRE FIGHTERS

Fire fighters should take the following steps to protect themselves from injury and death:

1. Follow all established policies and procedures.

Figure 15.4. (continued)

2. Wear and activate your PASS device at the scene of every emergency.

3. Wear the appropriate protective clothing and equipment (including your SCBA) at all incidents where hazardous atmospheres might be encountered.

4. Check your SCBA to assure that it is in working order and has been properly maintained.

5. Drink fluids frequently and be aware of signs of heat stress [NIOSH 1985, 1986].

ACKNOWLEDGMENTS

The principal contributors to this Alert were Ted A. Pettit, Tim R. Merinar, Michael A. Commodore, and Richard M. Ronk, Division of Safety Research, NIOSH. Please direct any comments, questions, or requests for additional information to the following:

Director
Division of Safety Research
National Institute for Occupational Safety
 and Health
1095 Willowdale Road
Morgantown, WV 26505–2888

Telephone, (304) 285–5894; or call
1–800–35–NIOSH (1–800–356–4674).

We greatly appreciate your assistance in protecting the lives of U.S. workers.

Linda Rosenthal

Linda Rosenstock, M.D., M.P.H.
Director, National Institute for
 Occupational Safety and Health
 Centers for Disease Control and
 Prevention

Injuries and Deaths of Fire Fighters

REFERENCES

ASME [1990a]. ASME standard A17.1–1990: Safety code for elevators and escalators (as amended), Section 211. New York, NY: American Society of Mechanical Engineers.

ASME [1990b]. ASME standard A17.3–1990: Safety code for existing elevators and escalators (as amended), Section 211. New York, NY: American Society of Mechanical Engineers.

CFR. Code of Federal regulations. Washington, DC: U.S. Government Printing Office, Office of the Federal Register.

IAFF [1994]. Line-of-duty death database. Washington, DC: International Association of Fire Fighters.

Morris GP, Brunacini N, Whaley L [1994]. Fire ground accountability: the Phoenix system. Fire Engineering *147*(4):45–61.

NFPA [1989]. NFPA 1404: standard for a fire department self-contained breathing apparatus program. Quincy, MA: National Fire Protection Association.

NFPA [1990]. NFPA 1561: standard on fire department incident management system. Quincy, MA: National Fire Protection Association.

NFPA [1992]. NFPA 1500: standard on fire department occupational safety and health program. Quincy, MA: National Fire Protection Association.

NIOSH [1985]. Occupational safety and health guidance manual for hazardous waste site activities. Cincinnati, OH: U.S. Department of Health and Human Services, Public

Figure 15.4. (continued)

Health Service, Centers for Disease Control, National Institute for Occupational Safety and Health, DHHS (NIOSH) Publication No. 85-115.

NIOSH [1986]. Occupational exposure to hot environments. Cincinnati, OH: U.S. Department of Health and Human Services, Public Health Service, Centers for Disease Control, National Institute for Occupational Safety and Health, DHHS (NIOSH) Publication No. 86-113.

NIOSH [1987]. NIOSH guide to industrial respiratory protection. Cincinnati, OH: U.S. Department of Health and Human Services, Public Health Service, Centers for Disease Control, National Institute for Occupational Safety and Health, DHHS (NIOSH) Publication No. 87-116.

NIOSH [1990]. Hazard evaluation and technical assistance report: International Association of Fire Fighters, Sedgwick County, Kansas. Morgantown, WV: U.S. Department of Health and Human Services, Public Health Service, Centers for Disease Control, National Institute for Occupational Safety and Health, NIOSH Report No. HETA 90-395-2121.

NIOSH [1993]. Fatal injuries to workers in the United States, 1980-1989: a decade of surveillance; national profile. Cincinnati, OH: U.S. Department of Health and Human Services,

Public Health Service, Centers for Disease Control and Prevention, National Institute for Occupational Safety and Health, DHHS (NIOSH) Publication No. 93-108.

NIOSH [1994a]. Hazard evaluation and technical assistance report: Memphis Fire Department, Memphis, TN. Morgantown, WV: U.S. Department of Health and Human Services, Public Health Service, Centers for Disease Control and Prevention, National Institute for Occupational Safety and Health, NIOSH Report No. HETA 94-0244-2431.

NIOSH [1994b]. National Traumatic Occupational Fatalities (NTOF) Surveillance System. Morgantown, WV: U.S. Department of Health and Human Services, Public Health Service, Centers for Disease Control and Prevention, National Institute for Occupational Safety and Health.

Washburn AE, LeBlanc PR, Fahy RF [1991]. Report on fire fighter fatalities 1990. NFPA J 85(4):46-58, 90-91.

Washburn AE, LeBlanc PR, Fahy RF [1992]. Report on 1991 fire fighter fatalities. NFPA J 86(4):40-54.

Washburn AE, LeBlanc PR, Fahy RF [1993]. Report on fire fighter fatalities in 1992. NFPA J 87(4):44-53, 68-70.

Figure 15.4. (continued)

NIOSH ALERT

Preventing Injuries and Deaths of Fire Fighters

> **WARNING!**
>
> Fire departments must review their safety programs and emergency operating procedures; failures to establish and follow these programs and procedures are resulting in injuries and deaths of fire fighters.

Fire departments should take the following precautions to protect fire fighters from injury and death:

- Establish and implement an incident management system with written standard operating procedures for all fire fighters. The system should include a well-coordinated approach to the emergency, accountability of all fire fighters, and provisions for their overall safety at the scene of the emergency.

- Develop and implement a written respirator maintenance program for all respiratory protective equipment used by fire fighters. Establish service and maintenance procedures and rigidly enforce them to provide respirators that are dependable and are constantly evaluated, tested, and maintained.

- Establish and implement a system to account for the location and function of all companies, units, and fire fighters at the scene of an emergency.

- Employ a buddy system whenever fire fighters wear self-contained breathing apparatus (SCBAs).

- Provide personal alert safety system (PASS) devices and ensure that fire fighters activate them when they are involved in fire fighting, rescue, or other hazardous duties.

- Encourage municipalities to review and amend their elevator and life safety codes to require fire fighter control for all elevators with a total travel distance greater than 25 feet.

- Guard against heat stress and other medical emergencies at the fire scene; provide cool water supplies, rest areas, and access to emergency medical personnel.

Fire fighters should take the following steps to protect themselves from injury and death:

- Follow all established policies and procedures.

- Wear and activate your PASS device at the scene of every emergency.

- Wear the appropriate protective clothing and equipment (including your SCBA) at *all* incidents where hazardous atmospheres might be encountered.

- Check your SCBA to assure that it is in working order and has been properly maintained.

- Drink fluids frequently and be aware of signs of heat stress.

Photo by Glen E. Ellman (printed with permission)

Fire fighters in protective gear at the scene of a fire.

Please tear out and post. Distribute copies to workers.

See back of sheet to order complete Alert.

Figure 15.4. (continued)

Call 1-800-35-NIOSH (1-800-356-4674) for additional information or for free single copies of the complete *NIOSH Alert: Request for Assistance in Preventing Injuries and Deaths of Fire Fighters* [DHHS (NIOSH) Publication No. 94-125].

U.S. DEPARTMENT OF HEALTH AND HUMAN SERVICES
Public Health Service
Centers for Disease Control and Prevention
National Institute for Occupational Safety and Health

Delivering on the Nation's promise:
Safety and health at work
for all people
Through prevention

DHHS (NIOSH) Publication No. 94-125

Figure 15.4. (continued)

Concept of the Personnel Accountability System

Unfortunately for the fire service, the personnel accountability system is a reactive rather than a proactive program. Statistics and reports indicate that a significant number of fireground fatalities are due to freelancing and a lack of accountability of crew members. Departments have been forced to develop personnel accountability systems because of litigation rather than developing them for the safety and health of their personnel.

Many departments have recognized the importance of developing and implementing a personnel accountability system and have done so from a proactive standpoint. For whatever reason—resistance to change, failure to recognize the significance of this system, interference with incident operations, or apathy—the implementation process has been very, very slow. Even departments with excellent personnel accountability systems have problems with implementation and usage at an incident scene. The fire service faces some serious problems, listed below, and does a poor job of addressing these problems on accountability.

- The service does not want to address the fact that there are significant risks at an emergency incident.
- The personnel accountability system slows down operations such as fire attack (the claim is that there is not enough time to implement the system at an incident).
- Personnel violate safety procedures, based upon their perception of risk.
- Personnel fail to utilize their PASS devices when entering a hazardous area.

The personnel accountability system is an excellent risk management tool that must be developed, implemented, and utilized in each department.

PASS Devices

An important part of personnel accountability is the use of personal alert safety systems (PASS). Developed in the early 1980s, these devices are known by such names as *personal alert devices, personal motion detectors,* or *personal locator devices*. The concept of these devices is that in the event that members become lost, disoriented, separated from their crews, or trapped, or that they experience any other type of emergency, this device will sound an alarm that will assist crews in locating them. The device will activate automatically if no motion is detected after a short period of time or can be activated manually by the wearer. Some newer PASS devices now incorporate a heat sensor as well, which detects thermal extremes, based on time and temperature. PASS devices were designed to be worn by the firefighters regardless of whether they were wearing SCBA. This ensures the safety of members if they are involved in nonstructural fires, such as on wildlands, or special operations.

Personal alert safety systems (PASS) have their designated function within the safety and health parameters. These devices are a significant part of the personal protective ensemble. These devices cannot be utilized in lieu of a personnel accountability system coupled with a solid incident management system. Due to complacency, apathy, system malfunctions, or for whatever reason, the usage (actual turning on of the devices at an incident) by personnel is very poor. Departments must ensure that these devices are used by personnel when operating at an incident scene.

THE PLAYERS

Each member has an assigned responsibility relating to the utilization of the personnel accountability system. In order for the system to function properly, everyone must play by the rules and contribute his or her share for the success of the system. Based upon each individual system, there may be personnel involved other than the ones mentioned.

Firefighters

Personnel must ensure they maintain crew integrity at all times during an incident. If one person leaves an area, all personnel leave; no one stays behind or is left alone. Depending on method of arrival, fire fighters must provide their nametag on the passport.

Company Officers

Company officers must track and account for all personnel assigned to them. The passport they provide must be accurate and include only the personnel entering a hazardous area. At the request of command or when a benchmark is reached, they must provide a correct "personnel accountability report" to the accountability officer.

Sector Officers

Personnel assigned as sector officers must be able to account for all crews in their sector. The sector officer must track the exact location of all crews and maintain the passports for the crews operating in the hazard area.

Command

The incident commander is responsible for designating an accountability officer, tracking the location of all crews, and ensuring that all companies arriving know the location for the accountability officer for entry and exit points.

STANDARD COMPONENTS OF THE PERSONNEL ACCOUNTABILITY SYSTEM

There are multiple components in a personnel accountability system, but several primary elements must be included in each personnel accountability system. The components are:

- Written standard operating procedures (SOPs)
- Training of personnel about the system components
- Hardware (e.g., helmet markers, tactical worksheets)
- Change/attitude
- Integration into the incident management system
- Development, implementation, and revision of system
- Noting of benchmarks
- Accountability officer
- Crew integrity
- Rapid-intervention crews
- Entry control and hazard zone

Each component plays an integral part of the process, as explained below.

Written Standard Operating Procedures (SOPs). As with any policy or procedure, the personnel accountability system must be in writing. The written policy details all the components of the system, providing examples and guidelines to ensure that personnel comprehend each concept and component of the system.

Personnel Accountability

Training of Personnel about the System Components. Most departmental policies are distributed, self-instructive, and self-implemented. It is imperative that a department utilize in-service training, company drills, and so on, as a method for introducing the personnel accountability system. In order to ensure that personnel understand the philosophy and intent of the personnel accountability system, classroom and practical training must be given.

Hardware (e.g., Helmet Markers, Tactical Worksheets). One of the vital parts of this system is the actual hardware. In order for the process to work, hardware such as helmet markers/identifiers, nametags, tactical worksheet, and status board have to be on scene and in use.

Change/Attitude. As with any new program, the department will have to ensure that personnel understand the value of the personnel accountability system. This will require a modification of attitude and understanding on behalf of the entire department.

Integration into the Incident Management System. As the personnel accountability system is developed and implemented, the department must ensure that the incident management system is revised to incorporate this new policy. A new procedure is being added to the incident management system that must be utilized by the incident commander at each incident scene.

Development, Implementation, and Revision of System. As with the risk management process, personnel accountability must be monitored for a period of time to determine whether any modifications need to be made. As the system is developed, implemented, and utilized, any necessary changes need to be noted and corrected during the revision process.

Noting of Benchmarks. During the actual operation of an incident, the incident commander needs to be aware of particular benchmarks, such as the time of implementation of the personnel accountability system (in five- or ten-minute intervals) and when a particular task is completed (such as fire extinguishment). These benchmarks will require an accountability check of all personnel operating in a hazardous area or zone. Other benchmarks requiring a personnel accountability report include:

Missing, lost, or trapped firefighters
A change from offensive to defensive operations
A significant event at the emergency scene
Completion of a primary or secondary search
Elapse of time intervals as dictated by policy
The request of the incident commander

Accountability Officer. In order for the personnel accountability system to function, an accountability officer will be needed to manage the process. It is very difficult for the incident commander to manage this task and manage the incident, too. This is where policy will dictate that a particular officer or company officer will assume this function.

Crew Integrity. Probably the most important reason for developing and implementing this system is to ensure that crews remain intact throughout their assignments at an incident scene. Nothing creates chaos faster at an incident scene than if a firefighter is missing or lost. Such a situation demands that an incident commander change the incident action plan to locate the persons rather than control the incident.

Rapid-Intervention Crews. During incident operations, the incident commander must provide dedicated personnel who are designated as rapid-intervention crews that will function as rescue crews if the need arises for personnel operating in a hazardous area. The rapid-intervention crew has no other assignment than to provide for the rescue of department personnel if the need arises. These personnel will wear appropriate protective clothing and equipment.

Entry Control and Hazard Zone. An important component of the personnel accountability system is that all personnel must enter into the process upon arrival at the incident scene. This allows the incident commander to enter crews into the system and provide assignments for them. The personnel accountability system provides control of crews entering the incident scene, regardless of their arrival sequence or their method of travel (e.g., pumper, ladder, personal vehicle). Personnel are not allowed to enter a hazard area without being entered into the system, thus preventing freelancing.

SYSTEM FEATURES

By employing a personnel accountability system, several significant and vital responsibilities of the management of emergency scene operations are accomplished. Some of these responsibilities include:

> Safety and health of all personnel operating at an incident scene
> Management of these personnel
> Location and function of assigned crews
> Proper utilization of resources (personnel and equipment)

VIRGINIA BEACH FIRE DEPARTMENT	SOP O 34
MPAS Guideline	07/01/93

MODIFIED PERSONNEL ACCOUNTABILITY SYSTEM (MPAS) GUIDELINE

GUIDELINE STATEMENT
The purpose of this guideline is to provide a safe, accurate, and efficient system of accounting for all fire personnel during any emergency incident. This system is designed to complement and work within the framework of the Virginia Beach Fire Department Incident Management Policy, but should not inhibit personnel from taking those emergency actions needed to avert personnel or civilian tragedy.

This guideline shall be used by all personnel for incidents in which the Fire Department has primary responsibility, and shall be used to account for Virginia Beach Fire Department personnel during incidents where other agencies have responsibility for the command function.

This guideline will change the way we conduct business. It is designed to track and account for all emergency personnel at an emergency incident. It complies with NFPA 1500 (Fire Department Occupational Health and Safety Program, 1992 edition) and NFPA 1561 (Fire Department Incident Mangement System, 1990 edition). The following objectives are critical to the success of this system:
1. All companies and personnel shall become a part of the accountability system and work for the Incident Manager, a sector officer, or a company officer. Freelancing (the preformance of a task without the knowledge of the supervising officer) is prohibited.
2. A minimum company size should be two or more members with a portable radio.
3. All companies should go in together, stay together, and come out together.
4. Each supervisor shall have a clear understanding of the position and function of each person assigned to them.

SYSTEM COMPONENTS
The Accountability System uses helmet shield, name tags, PASSPORTS, status boards, and make-up kits to account for and identify companies and individuals on the incident scene. All personnel are responsible for ensuring that the helmet shield and PASSPORTS always remain current (up-to-the-minute) and intact.
1. Helmet Shields:
Each helmet shall be equipped with magnetic strips permanently attached to both sides of the helmet. A second magnetic strip with a reflective helmet shield (E-14, L-9, etc.) shall be placed over the first strip. Personnel shall use only that helmet shield which shows their present assignment. Personnel assigned to brush, utility, or other vehicles housed at the same station shall retain the identity of that company's helmet shield.

If personnel leave an engine, ladder, etc., due to relief, leave, or other reason, the helmet shield for that unit stays with the unit. Extra helmet shields shall be kept on each unit.
2. Name Tags:
All members of the Virginia Beach Fire Department shall be issued four (4) plastic name tags (1/2" x 2") with velcro backing with should contain the last name and first initial of each firefighter or officer. The company officer shall keep one name tag of each firefighter assigned to him/her in a readily available location at the station should a full set of name tags be lost.

The name tag shall be placed on the PASSPORT (see below) of the unit that the firefighter or officer is currently assigned to ride. Personnel leaving a unit shall remove their name tag from the unit PASSPORT and place it under the brim of the helmet with their spare name tags.

Regional Team members for both Haz Mat and Technical Rescue Teams shall be provided with name tags for their response to incidents in Virginia Beach.

Figure 15.5. Virginia Beach Fire Department's Modified Personnel Accountability System Guideline.

VIRGINIA BEACH FIRE DEPARTMENT	SOP O 34
MPAS Guideline	07/01/93

The name tags shall be color-coded as follows:
Chief Officers: Black with white letters
Company Officiers: Red with white letters
Firefighters: Green with white letters
Volunteers: Yellow with black letters
Regional Teams: Gray with white letters
Blank Tags: White

3. PASSPORTS:
PASSPORTS are 2" × 5" strips of velcro attached back-to-back. A unit designation is etched into a plastic tag at the top and a removable unit tag is placed at the bottom. The front side of the PASSPORT is used to hold the name tags of current personnel riding the particular apparatus. The back side of the PASSPORT sticks to a velcro strip in the area of the officer/passenger side door on a Unit Status Board (see below). Name tags should be placed on the PASSPORT in the following order:
OFFICER OR FRONT SEAT PASSENGER
DRIVER (upside down)
JUMP SEATS
TILLER (upside down)

4. Status Boards:
Status Boards are used to hold multiple PASSPORTS during an incident. Space is available to write specific assignments given, the times in and out of hazardous areas, and other comments. There are two (2) types of status boards: Unit status boards and Command status boards.
  A. Unit status boards are 8 inches × 12 inches with velcro on both sides. They should be attached to the inside of the officer's side door on all fire apparatus. This board will hold the apparatus PASSPORT and can be used as a PASSPORT collection point during working incidents.
  B. Command status boards are 12 inches × 24 inches with velcro strips on outer sections of the board and a tactical worksheet in the middle section. These boards shall be kept in the Battalion vehicles, Support 1, and Tech 1. This board should be used by the Incident Manager or his/her designee to manage working incidents. It can also be used to manage larger numbers of PASSPORTS at specific points of entry, such as high-rise fires, technical rescues, hazardous materials incidents, etc.

MAKE-UP KITS
These kits shall be kept in the Battalion Chief vehicles, Support 1, and Tech 1. These kits are used when other agencies assist Virginia Beach or when companies of volunteers and/or off-duty personnel are formed at the incident scene. These BP kit pouches contain the following:
1. blank PASSPORTS
2. blank name tags (white in color)
3. markers for writing
4. spare sets of numbered hemet shields
5. sets of double hook or double loop velcro strips
6. incident site passes (see point-of-entry-control)
Personnel formed at the incident may not be assigned to a specific engine, ladder, tanker, or salvage company. When this happens, helmet shields with the prefix "TM" for TEAM, followed by a number, will signify the personnel formed at the scene (TM 4 or TM 8). Each team leader, chosen by the Incident Manager or his/her designee,

Figure 15.5. (continued)

VIRGINIA BEACH FIRE DEPARTMENT	SOP O 34
MPAS Guideline	07/01/93

should be provided with a portable radio and use the helmet shield to identify the team by radio.
Example:
"Team 10 to Command, salvage and overhaul of the 3rd floor is complete."

LEVELS OF ACCOUNTABILITY
The Accountability System shall operate at one of three levels during emergency incidents.
1. Level I:
   This level of accountability is established once units have arrived at or are assigned to a task at an emergency incident.
   The following shall take place during LEVEL I Accountability:
   A. The personnel and PASSPORTS of all staged companies will remain with their unit if not requested.
   B. When personnel are needed at the scene, the company officer shall bring the PASSPORT to the first arriving unit (collection point) and place it on the Unit Status Board.
   C. When companies are assigned a position remote from the collection point, the first unit at a remote entry point should be the collection point for that side or sector of the incident. The PASSPORTS and tracking of those companies should be delegated to a sector officer as soon as possible. Confirmation of a remote entry point shall be acknowledged by the Incident Manager.
2. Level II:
   This level of accountability is initiated during working incidents once Command has been established.
   The following should take place during LEVEL II Accountability:
   A. When a fire officer takes command of the incident, the Unit Status Board and PASSPORTS shall be brought to their vehicle if requested.
3. Level III:
   This level of accountability is initiated when the Incident Manager has established sectors and/or point-of-entry control.
   The following shall take place during LEVEL III Accountability:
   A. The Incident Manager shall have the PASSPORTS of all active companies at the Command Post.
   B. If Sectors are assigned, the PASSPORTS of those companies working within the specific sector are assigned to that Sector Officer. The Incident Manager should keep the removable unit tag for use at the Command Post.
   C. When desired, the Incident Manger can establish an Accountability Sector or Officer to take control of accounting for incident personnel.

PERSONNEL ACCOUNTABILITY REPORTS (PAR) AND TACTICAL BENCHMARKS
Just as an Incident Managment System has benchmarks (reference points) during an incident (life safety, fire control, and property conservation), the accountability system also has timely benchmarks when a roll call (PAR) of all personnel should be completed. PARs should be given face-to-face within the company or sector whenever possilbe, shall include the number of personnel, should be required in the following situations:
1. When the Incident Manager feels there is a need to confirm the location and assignment of any company at an incident, a roll call (PAR) including the number of personnel may be requested.

Figure 15.5. (continued)

VIRGINIA BEACH FIRE DEPARTMENT	SOP O 34
MPAS Guideline	07/01/93

2. At 20 minutes elapsed time of a working incident and every 20 minutes thereafter until Command advises the fire is under control or the incident stabilized. The dispatcher shall use the pre-alert tone and advise Command of the elapsed time. Example: (Pre-alert tones) "Dispatcher to Green Lakes Command, you have been operating on the scene for 20 minutes."
3. By any company completing a primary search. Company officers will ensure a PAR of their own company when they report a "Primary search complete, all clear and PAR 3."
4. Any report of a missing or trapped firefighter.
5. When an evacuation is ordered.
6. Any change from an offensive mode to a defensive mode.
7. Any sudden hazardous event at the incident (flash over, backdraft, explosion, or collapse).

MISSING OR TRAPPED PERSONNEL
An absent member of any company or team should automatically be presumed missing or trapped in the hazardous area until otherwise determined to be safe. The company officer shall immediately report any absent member to sector officers or Command.

A PAR shall be provided and all companies shall report immediately to their assigned sector officer or the Incident Commander, verifying no other lost or missing firefighters.

EVACUATION
When a hazardous condition is found at an incident requiring the immediate evacuation of personnel, all personnel shall remove themselves from the collapse zone and provide a PAR to the Incident Manager. The following procedures shall be followed:
1. The Incident Manager shall advise all units on the scene that an evacuation is taking place. The Tac Channel shall be used.
   Example: "Mill Landing Command to all units, Evacuate the structure!"
2. The Incident Manager shall have all apparatus drivers signal with rapid, continuous blasts of the air horn as a second means of alerting companies to leave the hazardous area.
3. The Incident Manager shall advise the dispatcher on the Fire Command Channel of the evacuation.
   Example: "Mill Landing Command to dispatcher, I've ordered an evacuation of the structure at 1675 Mill Landing Road."

POINT-OF-ENTRY CONTROL
Point-of-entry control shall be established at any incident where the Incident Manager feels the need for more stringent accountability. This might include high-rise fires, technical rescues, haz mat incidents, or other incidents which create the need for tighter control of access to a hazardous area.

PASSPORTS should remain with the designated sector officer at the "point-of-entry" to the hazardous area. A company shall give its PASSPORT to the sector officer when entering and collect their PASSPORT when leaving. Status boards shall only contain PASSPORTS of those companies in the hazardous area.

Companies exiting at a different location other than the original point of entry shall immediately notify their sector officer. The PASSPORT shall be retrieved.

When it becomes impossible to retrieve the PASSPORT immediately, and the company is reassigned to another sector, a Make-Up PASSPORT with spare name tags shall be assembled by the new sector officer. The original sector officer must be made aware of the change.

Figure 15.5. (continued)

Personnel Accountability

VIRGINIA BEACH FIRE DEPARTMENT	SOP O 34
MPAS Guideline	07/01/93

VOLUNTEERS
1. Those volunteers who are certified by the Battalion Chief of Training as active volunteers are eligible to receive name tags to be used on PASSPORTS.
2. When riding fire apparatus, volunteers will wear the helmet shields of the unit they are riding on. They should remove the helmet shields from the helmet and name tag from the PASSPORT when leaving a unit.
3. Those volunteers arriving at an incident in their private vehicles shall identify themselves to the Incident Manager before receiving an assignment.
4. Once assigned to a company at an incident, volunteers shall remain with that company until released by the Incident Manager.

GLOSSARY
Accountability—The responsibility of the Incident Manager and all personnel to account for each other at an emergency incident.
Accountability System—A guideline and procedure used to keep track of the location and assignment of personnel operating at an emergency incident.
Collection Point—The location where PASSPORTS are collected and placed on a status board. This can be at an engine, ladder, or Incident Manager's vehicle.
Command Status Board—A large (12" x 24") status board used by chief officers at emergency incidents. It contains the tactical worksheet and on both sides has velcro strips which are used to collect PASSPORTS during an emergency incident.
Company—A number of personnel including a supervisor who are assigned to an engine, ladder, or other apparatus.
Emergency Incident—A situation involving the use of fire department personnel and equipment to provide life safety, fire ot hazard control, and/or property conservation.
Freelancing—The performance of a task without the knowledge of the supervising officer.
Hazardous Area—An area normally found inside or closely surrounding a structure or subject to incident related hazards unless otherwise determined by the Incident Manager.
Helmet Shield—A reflective, magnetic strip which attaches to the fire helmet. This strip shall show the unit assignment of fire department personnel.
Incident Manager—The Fire Department member in overall command of an emergency incident.
Incident Site Pass—A laminated, numbered card used to allow specialized personnel such as VNG, Virginia Power, EPA, etc., into the hazardous area to perform a supervised function. Others such as governmental officials, the media, etc., shall be given a site pass when allowed on the emergency scene by the Incident Manager.
Make-up Kit—A kit used to create teams of personnel or to replace missing items at an emergency incident.
Make-up PASSPORT—A PASSPORT made up at an emergency incident when personnel have not been assigned to a company have been reassigned to a location remote to the original assignment.
Name Tag—A color-coded plastic tag, backed with velcro, that identifies personnel by last name and first initial.
PASSPORT—Strips of velcro attached back-to-back which are designed to hold the name tags of personnel assigned to a particular engine, ladder, etc. The PASSPORT can be attached to a staus board and is used for tracking companies and personnel at an emergency incident.

Figure 15.5. (continued)

VIRGINIA BEACH FIRE DEPARTMENT	SOP O 34
MPAS Guideline	07/01/93

Personnel—A person or persons, career or volunteer, who are certified by the fire department to perform the duties and responsibilities involved at an emergency incident.

Personnel Accountability Report (PAR)—A report given by supervisors to sector officers or the Incident Manager during an emergency that all personnel under his/her supervision are accounted for.

Point-Of-Entry Control—The establishing of a specific entry point or "gate" when the Incident Manager feels the need for more stringent control of accountability.

Remote Entry Point–A location distant to a PASSPORT collection point because of distance, hazard, or building construction. This point becomes another entrance to the hazardous area.

Roll Call–The procedure used by the Incident Manager to confirm that personnel working in a hazardous area are safely accounted for.

Sector—The organizational level having responsibility for either a geographic area or specific function at an emergency incident.

Shall—Indicates a mandatory requirement.

Should—A recommendation or that which is advised but not required.

Staging—A sector function and location where personnel and equipment are assembled near or at an incident scene to await instructions or assignments.

Supervisor—A Fire Department member who has supervisory authority and responsibility over other members.

Team—Personnel formed at an incident who have not been assigned to a specific company.

Unit—A name for a single fire vehicle.

Unit Status Board—A plastic board (8" × 12") placed on the passenger side door of all engines and ladder trucks which holds the passport of a particular company. This board can be used to collect multiple passports during an emergency incident.

Reviewed by _____   FIRE CHIEF _____

Figure 15.5. (continued)

All personnel assigned to the incident are to be accounted for before arrival, through such methods as tags describing company assignments by position, tactical worksheets, bar-code systems, or any other system that allows for the accountability of personnel. If a department allows for personnel to respond to a scene via private vehicle, the accountability system must provide a means of allowing these members to be incorporated into the process. This could include mutual-aid companies or other responding agencies that are part of the incident operation.

Everyone must be part of this process. If personnel are not trained and educated on the importance and the components of the system, it will be very difficult to ensure compliance. It is human nature to resist change, and the need for implementation of a personnel accountability system is an excellent example. The process can be easily destroyed if even one member chooses to ignore or fails to be part of this system. Failure to comply with the requirements of the process can lead to serious injury or death. A personnel accountability system is one of the most important risk management tools a fire department can develop and operate. Failure to use the system is evident in firefighter fatality studies or reports detailing these situations.

Conclusion

The personnel accountability system is an excellent risk management tool when properly utilized. Based upon the existence of numerous programs throughout the country, the procedures and technology to efficiently implement a successful system are available. The intent of this system is to track and account for personnel operating at an emergency incident. In the event that something happens to firefighters, they can then be located in a quick and definitive manner. The personnel accountability system has room to expand as technology increases. Technology will provide easier methods for tracking crew locations and locating lost firefighters.

PROFILE

Chapter 16: Incident Management System

MAJOR GOAL:

To integrate sound risk management principles into each component of the incident management system that has been developed, implemented, and been utilized by the organization

KEY POINTS:

- Determine whether your incident management system meets the requirements of a nationally recognized model for incident management.

- Determine whether the incident management system is utilized on all incidents, whether formally or informally.

- Determine when the incident management policy was last revised.

- Determine whether the system is flexible enough to be used for any type of emergency incident.

- Determine whether your incident management system allows for the integration of other responding agencies.

- Determine whether the organization's personnel accountability system has been integrated into the incident management system.

- Determine whether rapid intervention crews are utilized at all incidents as needed.

- Determine whether the rest and rehabilitation of personnel are routinely considered by the incident commander.

- Understand that proper incident management directly affects member health and safety.

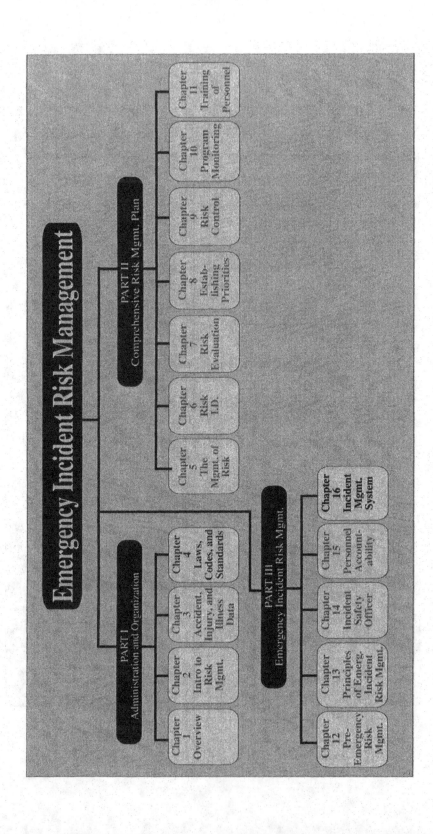

Chapter *16*

Incident Management System

INTRODUCTION

The Anytown Fire Department responds with an engine and ladder company, a heavy rescue unit, and two advanced-life-support (ALS) ambulances to a report of a vehicle accident on the northwest side of town. The engine company arrives to find a single vehicle accident with one car completely engulfed by fire. The engine company officer initiates an immediate fire attack and does not investigate to determine whether other vehicles are involved in this accident. Engine company personnel arrive wearing various stages of protective clothing, due to the fact they were dispatched to a "vehicle accident" from a station only a quarter of a mile from the incident. They begin to attack and extinguish the fire. The ladder company arrives and assists with the extinguishment. Neither officer assumes command of the situation because both are concentrating on attacking the fire. The squad officer and her personnel begin vehicle extrication when it is determined that two people are trapped in the vehicle. Levels of protective clothing vary for the squad members, due to the fact they were dispatched to a "vehicle accident." Both medic units arrive and pull up near the burning vehicle, with one unit parking on top of a handline being used for fire attack. This causes a sudden delay in extinguishing the fire and quite a bit of confusion. Personnel from both medic units walk around the incident scene, wearing no protective clothing (only station work uniforms) and having no equipment, and with no apparent assignment.

Police personnel inundate the scene because they were in pursuit of the occupants of the vehicle and are freely walking around the incident scene. The fire is somewhat contained, and extrication begins to free the trapped occupants. Finally, after 25 minutes on the scene, both patients are extricated and removed to the medic units. The fire has not yet been completely extinguished.

From this scenario, it is not difficult to identify the problems that were generated by a lack of management at this incident. Problem after problem developed, causing the overall problem to escalate and make this situation nothing more than an out-of-control mess. This scenario, with slight variations, occurs several times daily throughout this country. The types of equipment and number of personnel may vary, but the consistent factors are that without the use of an incident management system, the incident is inadequately managed, the risks to personnel are not controlled, and safety is nonexistent.

OBJECTIVES OF THE INCIDENT MANAGEMENT SYSTEM

The primary goal or objective of the incident management system is to manage personnel, ensuring effective service delivery, and to ensure the safety of our members at each incident scene, regardless of the size of the incident. An organization must establish and employ a system that allows for one individual to adequately command an emergency scene through a set of standard operating guidelines. This individual has the ultimate responsibility for the safety of all members operating at the emergency scene. The emergency scene may be a multivehicle accident involving trapped persons, a fire in a high-rise building, a hazardous-materials incident, treatment of a sick patient, or a room-and-contents fire. The intent is that whether the response involves one unit or multiple units from one or more organizations, an incident management system must be utilized by emergency response organizations.

An incident management system is an established policy detailing the functions, responsibilities, and operating procedures used to manage and control emergency incidents. This system may also be known as an incident command system or fireground command system. Emergency incidents present a myriad of hazards and risks to personnel. The use of an incident management system provides a systematic approach to identifying and evaluating risks and hazards to personnel, apparatus, and equipment at an emergency scene.

Due to the hazards faced during emergency operations, there are inherent risks already built into the process. Unfortunately, it is impossible to eliminate all risks. The objective is to ensure, though an incident management system, that the incident commander recognizes the risks, reduces the severity of the risks by implementing control measures (e.g., SOPs, protective clothing and equipment, personnel accountability), and ensures that safety is the primary operational goal. The safety of personnel must always be the incident commander's primary consideration, especially as the incident action plan is developed and implemented.

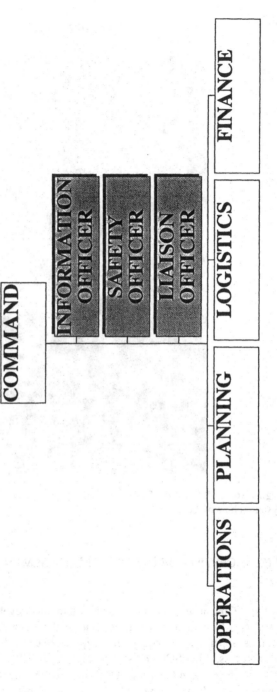

Figure 16.2. Organizational command structure for an incident management system.

Figure 16.3. The safety of personnel is the primary focus of an incident management system (photo by Martin Grube).

THE INCEPTION OF INCIDENT MANAGEMENT

In the 1970s, an incident command system was developed and implemented in California to combat brush fires. Due to the large-scale effort (personnel, resources, apparatus, and equipment) required to control and extinguish these types of incidents, FIRESCOPE was generated. FIRESCOPE stands for "Fire Resources of Southern California Organized for Potential Emergencies." This incident command system provides a means of managing emergency incidents regardless of the location within the state of California. Any department may initiate this system as the incident escalates due to the response of mutual-aid departments. Each organization that participates in the incident knows the organizational structure, which allows the incident to be effectively managed and controlled.

In the late 1970s, Fire Chief Alan V. Brunacini developed and implemented the Fireground Command System in the Phoenix Fire Department. Designing the Fireground Command System from procedures used by the Los Angeles City Fire Department to control high-rise fires, Chief Brunacini produced an everyday system for the Phoenix Fire Department to manage all firefighting situations and any other type of emergency, beginning with a single unit response and escalating up to multiple alarms. The Fireground Command System contains standard functions and procedures that combine to develop an effective incident organization. One of the primary values of the Fireground Command System is safety. The philosophy instilled by Chief Brunacini is that we can effectively manage and control an incident, yet provide for the safety, health, and welfare of personnel.

Chief Brunacini, through the Fireground Command System, challenged the fire service to look at the methods used to control fireground operations. Although there are inherent risks that come with the territory of fireground operations, the intent of the system is to evaluate and manage those risks in a calculated manner. By combining risk management, safety procedures, a functional command structure, and a competent incident commander, you ensure a successful, safe operation.

The next development in the incident management process came in the late 1980s from the National Fire Academy in the form of the Incident Command System. The Incident Command System incorporates components that can be used to control emergencies such as multicasualty, hazardous materials, and high-rise incidents.

In May 1990, at the NFPA Annual Meeting in San Antonio, Texas, NFPA adopted NFPA 1561, *Standard on Fire Department Incident Management System.* Developed by the NFPA Fire Service Occupational Safety and Health Technical Committee, this standard provided a generic model for an incident management system incorporating the primary components needed to develop and implement this system.

In 1993, the National Fire Service Incident Management Consortium developed a model system that combines or merges the components of FIRESCOPE, the Fireground Command System, the Incident Command System, and NFPA 1561. The first document produced by this group is an incident management system model for structural firefighting. Other documents are evolving that will address model incident management systems for incidents involving emergency medical operations, hazardous materials and special operations, and high-rise incidents.

With a myriad of incident management systems to utilize, emergency service organizations are at a strict disadvantage in failing to employ a command structure. Not only do customers suffer, but, more importantly, personnel suffer the consequences in the form of fatalities and serious injuries.

Incorporating Risk Management into the Incident Management System

Regardless of the situation encountered, the first arriving officer must evaluate conditions, which includes identifying risks to personnel who will be operating at the incident. This evaluation process must continue for the duration of the incident. As the strategies are developed by the incident commander, risk management and the safety of personnel must be included and continuously evaluated to ensure the health and welfare of personnel. There is always the option of changing the strategy, based upon future conditions or hazards that occur or are encountered.

If an engine company encounters a condition in which a person is trapped in a trench collapse or cave-in upon arrival, the officer must evaluate or size up the situation, incorporating the condition of the trapped person, the strategy options that must be made to safely extricate the person from the trench, the risks to personnel who must extricate the trapped person, and the tactics that will be employed to accomplish this extrication. Based upon the officer's training, he should realize that he cannot immediately commit personnel to rescue the trapped person. This decision is supported by OSHA regulations, which detail the proper procedures for safely and successfully completing such an operation. Decisions such as these are not easily made and are greatly influenced by human emotion and the desire to immediately commit personnel to remove the individual from the confines of the trench, totally disregarding the risks. The officer knows he must immediately request properly trained personnel and specialized equipment to accomplish the rescue of the trapped person.

Upon arrival of the first unit, the initial step of the officer will be the establishment of the department's incident management system. A continuous evaluation of the incident scene must be made and strategies and tactics revised when any condition changes, such as the arrival of additional personnel and equipment. With these revisions, the safety of personnel must be paramount. As the incident escalates, the engine company officer may transfer command to a battalion officer. With the transfer of command, the officers confer about what was observed and happening upon arrival of the first units, the decisions that have been made up to this point, and what will be needed to control the incident from this point. Regardless of who is incident commander, the focus remains on the safety and health of personnel.

These are just two examples of types of situation encountered, but they are similar from the standpoint that the primary focus of an incident is on the safety of personnel, effectively managing the risks, and controlling the incident.

Risk Philosophy

Identifying, evaluating, managing, and controlling the risks at an emergency incident requires the philosophy that an organization must ingrain in all organi-

Incident Management System 253

zational personnel. Based upon the nature of our business or the service that we provide to our customers, there will be risks that are encountered at each incident. Customers usually do not call us because they are having a terrific day or want us to stop by for a visit and chat. The reality is that it will be impossible to eliminate all risks that are confronted at an incident. The deep, dark secret is how we reduce the severity of these risks. Again, this is done by identifying what the risks are at an incident scene, evaluating them based upon degree of severity, and controlling them by use of an incident management system and proper protective clothing and equipment. The risks that cannot be completely eliminated will have to be managed by the incident commander through the use of an effective strategy and justifiable tactical operations.

An effective risk analysis begins with the arrival of the first officer on scene. An evaluation system must be used to develop the incident action plan and forecast the outcome. A standard evaluation system will provide an immediate initial risk analysis as well as a size-up of the situation. This size-up will generate a strategy, and the officer will assign tactical objectives to be used to control this incident. If an officer fails to employ the evaluation system, then the incident will begin to control the incident commander rather than the incident commander controlling the incident.

Incident Risk Assessment

As a department develops a risk management plan for incident operations, it would be futile to write policy identifying all the risks, developing control measures for these risks, and delineating what is acceptable risk. The evaluation of risk is an assignment that the incident commander is designated. The incident commander must perform a risk analysis to determine what hazards are present, what the risks to personnel are, based upon the incident action plan, how the risks can be eliminated or reduced, the chances that something may go wrong, and the benefits gained, based upon the strategy being employed.

The incident commander has the responsibility for employing risk management principles at an incident. All personnel must support this concept and utilize these principles, based upon their involvement. Just as safety is a value within an organization, risk management must also be accepted as a value.

Incident commanders must examine the acceptable levels of risk as the risks directly relate to the capability to save lives or property. The risk to personnel must be evaluated in proportion to the ability to save the lives of civilians and protect the lives of personnel. Chapter 6 of NFPA 1500 incorporates risk management verbiage for emergency operations. By incorporating risk management into the incident management system, the basis of emergency incident risk management is built on the following:

Regular assessment of the conditions
Essential decision making

Tactical design
Routine evaluation and modification
Procedural command and management

Providing a regular means of evaluating incident scenes ensures for a standard outcome. Responsible strategies plus competently chosen tactical objectives ensure the safety and welfare of personnel. As an incident continues, the incident action plan must be routinely evaluated and modified when necessary. Most importantly, risk management is integrated into the incident management system, which provides a means of procedurally managing and controlling incidents.

The risk management concept can be defined utilizing the following principles:

We will risk a lot to save a lot.
We will risk a little to save a little.
We will risk nothing to try to save what is already lost.

This is an important philosophy that we must incorporate into the incident management system that the incident commander must use at each incident. Each officer and each firefighter must be trained to evaluate risks in this manner. These principles inform a decision-making process that will determine whether the strategy is an offensive operation or a defensive operation.

Risking a Lot

It is very easy to look at situations portrayed in pictures and videos showing firefighters at risk in performing rescue operations or conducting interior operations, and say that they are being unsafe. In most cases, we know little about the situation and conditions encountered by these personnel. The criticism may come in the form of lack or improper use of protective clothing and equipment. Under that assumption, it is assumed that the rest of the operation is conducted in the same manner: unsafe. The point is that conditions may exist in which a quick decision was made to effect a rescue in an unsafe environment, for example. The firefighter enters a "well-involved" room to rescue a victim, unsure whether a flashover will occur. Any firefighter worth his or her weight in salt would not disagree with this action. Personnel must ensure that they are fully protected when undertaking a rescue or search for possible victims. They may not have time to confer with their officer to ask for permission to "go for it." This situation is a calculated risk that the firefighters took, risking a lot to save a lot.

In another example, an incident commander arrives at an incident scene, taking command from the first arriving company officer. Crews are operating several handlines in the building, conducting a primary search and firefighting operations.

Incident Management System

Based upon the incident commander's observations and reports from interior crews, the interior conditions are quickly deteriorating. Realizing that the survival rate for any victims is minimal and that the risk to interior crews is increasing, the incident commander orders personnel from the building. This is a difficult situation for the incident commander, but the risk to personnel outweighs the benefits of leaving them in the building and having them become trapped or lost.

Risking a Little

The second principle is: "We will risk a little to save a little." Once the life hazard has been removed, firefighting efforts are turned to preserving saveable property. In extinguishing a fire, the closer we can get to the fire, the easier and the faster it can be made to go out. This saves a great many buildings and structures, but it has also killed many firefighters. The incident commander must perform a risk/benefit analysis of the situation that will assist in the decision-making process to determine the strategy of interior operations. As we have discussed, the incident commander uses reports from interior crews, information from the incident safety officer, and observation of conditions to monitor the operation constantly in the event that the strategy does not work out as planned. Upon indications that something is wrong or is beginning to go wrong, the incident commander can quickly change strategies and move to a defensive operation.

Risking Nothing

A defensive operation is a situation that employs the principle "We will risk nothing to save nothing or to try to save what is already lost." Abandoned buildings provide a prime example of the use of this principle. There are numerous scenarios detailing firefighter fatalities during interior operations at abandoned structures. Because of structural collapse, personnel becoming lost or separated from their company, or fire conditions, these firefighters were killed attempting to save property that has little or no value and that eventually will be razed. An incident commander must recognize that defensive operations are implemented when it is determined that there is nothing that can be saved. Personnel must be accounted for and given an assignment away from the structure. This prevents personnel from being tempted into going into the building for whatever reason.

Toolbox for Evaluating Incident Risk

As an incident commander conducts an evaluation of an incident scene to determine the strategy and tactical objectives and to measure the level of risk to personnel, there are some tools available to help with the process. These are instruments that the incident commander can use throughout the incident to effectively manage the incident and ensure the safety and welfare of all assigned personnel.

The components are as follows.

- Performance of a risk analysis by personnel at each incident, regardless of severity
- Multiple strategic plans
- Use of a recognized incident management system
- Competent training and education of all personnel in incident management operations
- Cynical approach to shifting situations
- Mandatory use of full protective clothing and equipment, including SCBA and facepiece
- Standard communication procedures that complement the IMS personnel accountability
- The transmission of elapsed-time notifications to the incident commander
- Available rapid-intervention crews for use in the event of an emergency
- Use of an incident safety officer and a safety action plan
- Providing of resources as necessary
- An aggressive "rehab" sector
- Periodic reevaluation of the conditions of the incident
- Use of the post-incident analysis and of lessons learned from past incidents

Each of these plays a critical role in the successful outcome of an incident. Each incident may require the use of some or all of these tools. It is the incident commander's prerogative to use whatever is needed to manage, control, and mitigate the hazards at the incident scene.

Performance of a Risk Analysis by Personnel at Each Incident, Regardless of Severity. Each firefighter, officer, and command officer must always perform a risk evaluation during an incident operation. This must be part of the initial size-up process and be performed continuously during the incident. This is especially important for the incident commander, who is responsible for the safety and welfare of all personnel operating at the incident scene.

Multiple Strategic Plans. The incident commander forecasts the outcome of the incident, using an effective incident management system, which incorporates

Incident Management System 257

risk management, safety, and an incident action plan. Forecasting allows the incident commander to develop contingency plans in the event current strategies are unsuccessful.

Use of a Recognized Incident Management System. In order to effectively manage emergency operations and to provide for the safety and health of personnel operating at emergency operations, a department must develop, implement, and use an incident management system. This incident management system must be detailed in writing so that personnel can be properly trained in the utilization of this policy.

Competent Training and Education of All Personnel in an Incident Management Operation. Any policy is ineffective if not implemented or properly used. Members must be trained on the components of the incident management system and how it is implemented for use in their organization, from the smallest incident to the largest incidents. The incident scene is not the place for training on the use of this system.

Cynical Approach to Shifting Situations. An incident commander "must plan for the worst and hope for the best" from the standpoint of managing emergency incidents. Forecasting plays a tremendous role in maintaining a pessimistic outlook on an incident.

Mandatory Use of Full Protective Clothing and Equipment, Including SCBA and Facepiece. Personnel must understand their part in the risk management process of incident scene management. Through policy and procedures, a department mandates the level of protective clothing to be worn during incident scene operations. When this becomes an issue of compliance, this distracts from the management of incidents and jeopardizes the safety and health of personnel. Policy is written to ensure the safety of personnel operating at incidents, based upon the risk assessment by the incident commander.

Standard Communication Procedures That Complement the Incident Management System. An organization must utilize a communication process that allows the incident management system to develop, to provide continuous feedback of conditions, and to funnel pertinent information to command. This communication should be in the form of clear, concise text. Organizational policy will assist with the development of effective and controlled incident scene communications.

Personnel Accountability. Regardless of the incident, maintaining control and location of personnel is foremost. The personnel accountability system must be integrated into the incident management system to reduce the chances of freelancing and the ability to effectively track and move personnel operating at an emergency incident.

The Transmission of Elapsed-Time Notification to the Incident Commander. A major element in firefighter safety on the incident site is the element of time. Many safety factors change dramatically as time goes along. Having a system that reminds the incident commander of elapsed time (generally in ten-minute increments) creates a safer and more effective situation.

Available Rapid-Intervention Crews for Use in the Event of an Emergency. As personnel are assigned operations in a hazardous area at an incident scene, the incident management system must provide assigned personnel for the rescue of these firefighters or other workers in the event that an emergency occurs. Rapid intervention crews (RIC) or firefighters assistance and search teams (FAST) are equipped with the proper protective clothing and equipment to enact a rescue of personnel should the need arise. These personnel are dedicated solely to this purpose while crews are operating at risk.

Use of an Incident Safety Officer and a Safety Action Plan. Though the incident commander is responsible for the safety of all personnel at an emergency incident, the appointment or utilization of an incident safety officer will strengthen this obligation. The incident safety officer will develop and operate a safety action plan that is integrated with the incident commander's incident action plan.

Providing of Resources as Necessary. As an incident continues to escalate, the incident commander must be aware of available resources that will additionally support the operations. This may be in the form of a backhoe and operator for a trench rescue, additional truck companies to augment defensive operations, or extra extrication tools at the scene of a multivehicle incident.

An Aggressive "Rehab" Sector. One important aspect of firefighter safety is maintainance of the well-being of personnel operating at an emergency incident. The "rehab" sector is an area that allows for the medical monitoring, rest, nourishment, safety, and serenity of personnel away from the operations of the incident. Weather extremes also affect the conditions of personnel, and the "rehab" sector must be established to provide for the physical welfare of these personnel.

Periodic Reevaluation of the Conditions of the Incident. Routinely, the incident commander must reevaluate the conditions at the emergency incident. This monitoring process, whether prompted by dispatch or other method, surveys hazardous situations, the condition of personnel, location of apparatus and equipment, and the effectiveness of the incident action plan and a need to change.

Incident Management System

Figure 16.4. "Rehab" is a crucial part of incident scene safety (photo by Martin Grube).

Use of the Post-Incident Analysis and Lessons Learned from Past Incidents. Once an incident has been terminated, an organization must critique the incident and discuss problems encountered, what worked well and what needs improvement, and any policy or procedures that need to be reviewed and revised for the betterment of the organization. If safety problems and other issues are not addressed, the same mistakes may be made at the next incident. The intent of the post-incident analysis is to learn from our mistakes, not to point blame for them.

CONCLUSION

An incident management system is an effective tool for managing personnel and controlling operations at an emergency incident. Failure to use an incident management system greatly jeopardizes the safety, health, and welfare of personnel. There are many components that can be utilized to make an effective incident management system. The incident commander must incorporate these components into the incident action plan to ensure a successful outcome of the emergency.

A department's incident management system must be able to function for a variety of incidents:

Fire suppression
Emergency medical operations
Hazardous-materials incidents
Special operations (confined space rescue, technical rescue, etc.)
Any other emergency situation

Also, the incident management system must be able to expand as an incident grows, such as a high-rise fire, wildland fire, or mass-casualty incident. Other organizations, such as mutual-aid companies, law enforcement personnel, and public works, will need to be integrated into this system. These are situations that will greatly test the system and demonstrate the real need for an incident management system.

Safety is one of the primary components of the incident management system. An incident commander must maintain this objective, from the initial dispatch of an incident until all personnel are back in quarters. Our goal is to have all personnel return from an incident in the same condition they were in when they got there. The incident commander is responsible for the safety of all personnel operating at an incident scene. This responsibility is delegated to all officers ensuring the safety of their personnel. Also, all members must recognize they have a responsibility to monitor their own safety as well as that of fellow personnel.

PROFILE

Chapter 17: Post-Incident Analysis

MAJOR GOAL:

To provide an opportunity to thoroughly review the overall operations of an actual incident, and to explain that the purpose of the post-incident analysis is to improve operations, the health and safety of personnel, and the quality of service to the customers at an incident

KEY POINTS:

- Develop written procedures that define the process to be followed when critiquing an incident.

- Improve the quality of service delivered to customers.

- Understand that the purpose is to identify positive actions and needed improvements.

- Understand the need for learning from any mistakes made in order to avoid making them again.

- Understand that the small mistakes in time can turn into large mistakes.

- Understand that the Incident Safety Officer and the Health and Safety Officer play an important role in the post-incident analysis process.

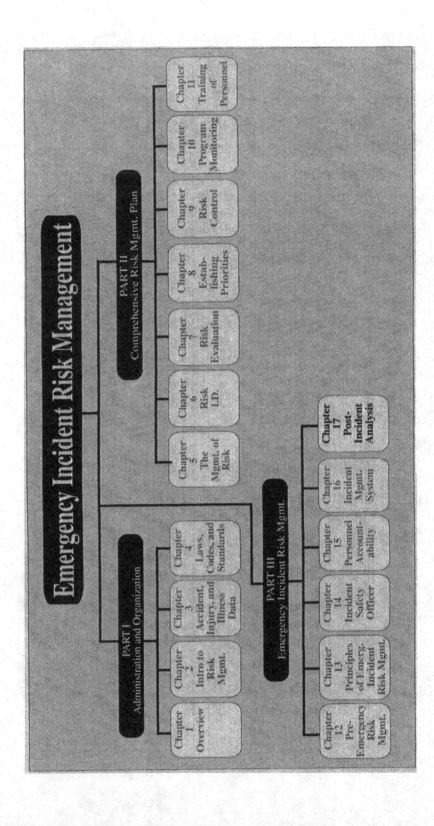

Chapter *17*

Post-Incident Analysis

INTRODUCTION

After an emergency incident has been completed, a department must systematically evaluate and review the operations that transpired. This process is commonly known as a critique, post-incident critique, or post-incident analysis. For the purposes of this book, the process will be known as and identified as the post-incident analysis. All of the significant players involved in the incident, including the incident commander, sector officers, company officer, and incident safety officer, need to be included as presenters.

The post-incident analysis includes a basic evaluation of the circumstances present, the operations, and the effect of the circumstances and operations relating to the safety of the personnel present at the incident.

NFPA 1500, *Standard on Fire Department Occupational Safety and Health Program*, Paragraph 6-8.1, states that the fire department should develop criteria and standard operating procedures (SOPs) for a standardized post-incident critique or analysis of significant incidents or those that involved serious injury or fatality of a firefighter. A significant incident may include a multiple-alarm fire, responding on or requesting mutual aid, or an extended incident. Each department must determine the criteria for instituting a post-incident analysis. The important point is that a review process is conducted to evaluate the operations that worked well and the procedures that need to be modified.

For most of us, it is very difficult to effectively evaluate ourselves at an incident when operations and outcome have not gone well. In order for the organization to benefit from its mistakes, the post-incident analysis is an avenue necessary to effect change through standard operating procedures, policy, and/or responsibilities.

The post-incident analysis must include the safety issues or concerns that occurred at the incident, both positive and negative. The incident safety officer must present the significant events that impacted the incident from a safety standpoint. If all operations worked well, personnel performed within the guidelines of the incident management system and the personnel accountability system as required, personnel wore proper protective clothing and equipment, and no accidents or injuries resulted from incident operations, then the incident safety officer should acknowledge these positive actions. The same is true for operations or actions that take place when department SOPs or department policy are not observed.

Should a responder fatality occur at an emergency incident, the impact of this event will be devastating enough. A preliminary report of what happened should be reported as soon as possible to eliminate or squelch rumors and to state exactly what occurred. If a significant injury occurs, there should be a discussion of what happened and what can be done to prevent recurrence, from an objective standpoint not a personal one. Remember, this is not done to embarrass, make fun of, or point fingers at a particular person or persons. The incident safety officer must identify what happened and how an organization can learn from this process to improve organizational operations.

BENEFITS AND COMPONENTS OF THE POST-INCIDENT ANALYSIS

The post-incident analysis must be regarded as a process that reviews the operations at the incident and how they may be improved. Benefits for instituting a post-incident analysis include the following:

- A complete systematic account of an incident and an evaluation of the effectiveness of department procedures
- Evaluation of company response times
- Evaluation of overall jurisdiction-wide protection, based upon the needs of the incident
- Review of effectiveness of tools and equipment
- Evaluation of safety practices and procedures
- Assessment of department training needs
- Assessment of effectiveness of department operations with outside agencies

In order to provide a systematic approach, each company officer, each sector officer, and the incident commander must detail:

- What they encountered upon arrival
- The initial strategy and tactics

The results of the strategy and tactics
Any obstacles that were encountered
What worked well and why
Recommendations for improvement

The post-incident analysis should be conducted as soon as all personnel complete their fact sheets and return them to the moderator.

A time frame should be attached to the post-incident critique or analysis to prevent the process from becoming nonproductive. The post-incident analysis should be moderated by a department chief officer other than the incident commander, so the process remains positive and less prone to conflict, and continues to be a productive learning experience.

The format should be constant to ensure that everyone knows and understands the rules. The format should allow first arriving officers to detail their experiences and strategies and then continue with other assignments as the incident expands. Issues and stages that need to be addressed are:

Introduction
Building design
Dispatch and response
Scene operations
Communications
Resource functions
Incident safety officer/safety sector
Accountability
Investigations
Conclusion

Members who had an assignment or part in this process must provide information as to their actions. They must be part of the post-incident analysis to ensure that all operations affecting the incident are presented.

SAFETY AND HEALTH ISSUES

The incident safety officer must identify operational issues to discuss during the post-incident analysis. The operational issues should focus on, but not be limited to:

Personal protective equipment
Accountability
Personnel safety and health concerns

VIRGINIA BEACH FIRE DEPARTMENT	SOP O 19
Incident Command Policy	05/01/91

POST INCIDENT ANALYSIS

DEFINITION

Post Incident Analysis (PIA) is the reconstruction of an incident to assess the chain of events that took place the methods used to control the incident and how the actions of emergency personnel contributed to the eventual outcome.

PURPOSE

The main purpose of the PIA is to reinforce personnel actions and departmental procedures that are effective and to give management insight into how effectiveness of the department's operations can be improved.

BENEFITS

- Provides a comprehensive analytical record of an incident from which to evaluate departmental procedures.
- Assessment of response times and company response areas under actual conditions.
- Assessment of the effects of additional equipment/manpower requests on over-all city wide protection and the effectiveness of back-fill procedures.
- Assessment of tools and equipment.
- Assessment of safety practices and related procedures.
- Assessment of training needs for department personnel.
- Assessment of the department's working relationship with outside agencies and other city departments.

PROCESS

- A formal PIA will be conducted for all incidents requiring additional heavy fire apparatus (pumpers and ladders) beyond a level of three pumpers and one ladder. A PIA may be requested of the first arriving units in the event of a fatality at the incident scene. An informal PIA can be conducted for any other incident at the discretion of the Incident Commander.
- A formal PIA process will be initiated by the Incident Commander.
- All incident commanders, company officers and non-fire department sector officers participating at the incident will complete a PIA fact sheet. These sheets are to be completed as soon as possible after the incident and submitted to the officer initiating the PIA process.
- Once all PIA fact sheets have been received and reviewed, the PIA will be scheduled and the participants notified in writing.
- The Incident Commander will lead the PIA following the structure of the PIA summary sheet. Discussion should include those areas on the summary sheet but may extend beyond at the discretion of the PIA leader or as the incident dictates.
- The fact sheets and the summary sheet will not be public documents. They will be internal worksheets for investigative information and department evaluation. The fact sheets are <u>NOT</u> to be attached to fire reports nor should any reference be made to the fact sheet on the fire report.

Figure 17.1 The Virginia Beach Fire Department Post Incident Analysis Standard Operating Procedure.

```
                POST INCIDENT ANALYSIS
                       SUMMARY

DATE _____       TIME OF ALARM _____
ADDRESS _____
TYPE OF INCIDENT _____

SITUATION UPON ARRIVAL OF FIRST UNITS: (INCLUDE A BRIEF DESCRIPTION
OF THE SITUATION ENCOUNTERED BY THE FIRST UNIT(S) ARRIVING ON THE
SCENE. THE TYPE OF UNITS AND MANPOWER ON UNITS SHOULD BE LISTED.)
_____
_____
_____
_____
_____
_____
_____
_____


FINAL OUTCOME OF INCIDENT: (LIST EXTENT OF DAMAGE AND CASUALTIES.
ALSO INCLUDE DAMAGE TO FIRE EQUIPMENT AND EMERGENCY PERSONNEL
CASUALTIES.)
_____
_____
_____
_____
_____
_____
_____


EQUIPMENT COMMITTED TO INCIDENT: (LIST MANPOWER AND UNITS COMMIT-
TED TO THE INCIDENT. INCLUDE PARTICIPATING VOLUNTEERS AND PAID OFF
DUTY PERSONNEL THAT RESPONDED IN PRIVATE AUTOMOBILES.)
_____
_____
_____
_____
_____
_____
_____
_____


EQUIPMENT AND MANPOWER NOT COMMITTED TO THE INCIDENT: (LIST THE
STATIONS LEFT EMPTY AND THOSE STATIONS BACKFILLED BY WHAT APPARA-
TUS.)
_____
_____
_____
_____
```

Figure 17.1. (continued)

STRATEGY: (LIST THE INCIDENT COMMAND STRATEGIES CHOSEN. INCIDENT COMMANDERS SHOULD DESCRIBE THEIR BASIC PLAN TO ADDRESS THE PRIORITIES OF THE INCIDENT AT THE TIME THEY BECAME THE INCIDENT COMMANDER.)

FIRST IN UNIT(S): _____

A. GENERAL STRATEGY _____
_____
_____

B. RESULTS _____
_____
_____
_____

FIRST INCIDENT COMMANDER (NAME): _____

A. GENERAL STRATEGY _____
_____
_____

B. RESULTS _____
_____
_____
_____

SECOND INCIDENT COMMANDER (NAME): _____

A. GENERAL STRATEGY _____
_____
_____

B. RESULTS _____
_____
_____
_____

THIRD INCIDENT COMMANDER (NAME): _____

A. GENERAL STRATEGY _____
_____
_____

B. RESULTS _____
_____
_____
_____

COMMON OBSTACLES (LIST THOSE PROBLEMS ENCOUNTERED BY MORE THAN ONE CREW OR IC THAT MAY INDICATE A NEED TO REVIEW DEPARTMENT PROCEDURES OR TRAINING.)

_____
_____
_____

Figure 17.1. (continued)

RECOMMENDATIONS (LIST ANY RECOMMENDATIONS FOR CORRECTION OR REDUCTION OF THESE OBSTACLES.) _____
_____
_____
_____
_____

WHAT OPERATIONS WORKED WELL? WHY? (LOOK AT STRATEGIES AND RESULTS, NOT ONLY AT THE IC LEVEL BUT AT THE SECTOR LEVEL IF APPROPRIATE. THAT HELPS REINFORCE PROCEDURES/TACTICS THAT WERE SUCCESSFUL SO THEY MAY BE APPLIED TO SIMILAR SITUATIONS IN THE FUTURE.)
_____
_____
_____
_____
_____
_____

INCIDENT COMMAND ORGANIZATIONAL CHART: (DRAW IN LINES OF AUTHORITY AND RESPONSIBILITY AND IDENTIFY SPAN OF CONTROL. THIS ALLOWS FOR A MORE FORMAL REVIEW OF THE IC PROCESS IN ORDER TO IDENTIFY THE POSITIVE ASPECTS AND CORRECT DEFICIENCIES.)

SUMMARY: (THE SUMMARY SHOULD BE WRITTEN BY THE HIGHEST RANKING IC AT THE INCIDENT.) _____
_____
_____
_____
_____
_____
_____

ADDITIONAL NOTES: (THE ORIGINAL OF THIS PIA SUMMARY SHOULD BE KEPT BY THE OFFICER WHO COMPLETES THE SUMMARY SECTION. A COPY OF THE FORM WILL BE SENT TO THE INCIDENT COMMANDER. THE INFORMATION IN THIS PIA SUMMARY SHOULD HAVE BEEN DISCUSSED DURING THE CRITIQUE.)
_____
_____
_____
_____
_____

Figure 17.1. (continued)

This identification process is intended to evaluate the department operations overall, not to concentrate on personnel problems or issues. Most importantly, the incident safety officer has to maintain a positive attitude and be proactive through the post-incident analysis.

Improving Incident Safety

The post-incident analysis will identify operations and performance that must be changed or revised and that relate to the safety and welfare of department personnel. These changes or revisions will come in the form of standard operating procedures (SOPs), policy, and training and education. Once these issues have been identified, a plan must be developed to make the suggested changes. This plan includes reference to the changes or revisions that need to take place, who is responsible, the dates the changes will be made, when they will become effective, and any other details that are necessary.

Emergency Evaluation and Analysis

Incidents, such as accidents and injuries, affecting personnel at emergency scenes may be investigated by the company officer, the employee's supervisor, or the incident safety officer. The health and safety officer and/or the incident safety officer involvement will depend on the nature and severity of the incident.

INTERFACING WITH THE INCIDENT SAFETY OFFICER

As program manager of the safety and health program, the health and safety officer must assist the incident safety officer as needed during an investigation or analysis of the incident.

Investigations

The nature or severity of the incident will dictate the degree of involvement of the health and safety officer. If a firefighter fatality occurs, the health and safety officer will be in charge of the investigation, based on time commitment and the magnitude of this process. The incident safety officer may intitiate the investigation. The health and safety officer will be placed in charge of the investigation upon arrival or at a designated time.

The health and safety officer may assist with the investigation if an additional resource is needed for any reason. Also, the health and safety officer may assist to ensure that the proper documentation is completed. For example, if additional information needs to be obtained from witnesses or department personnel, such as statements or interviews, the health and safety officer can do this. If protective clothing and equipment need to be impounded for testing, the health and safety officer would assist with this process.

The health and safety officer may be the resource for having the investigation report finalized in a standard report form. If the report needs to be properly formatted, with additional information from outside agencies or test results, the health and safety officer may be the focal point of this process.

Department Procedures

The investigation report may have recommendations for revision or development of procedures pertaining to departmental operations. The investigation report may indicate that procedures be developed based on actions at an incident. Based upon observations by the incident safety officer at an incident scene or accident scene, a recommendation for policy development or revision of current policy may be part of the report summary.

Training and Education

The recommendations may indicate deficiencies or inadequacies in department operations. If they relate to training and education needs, the health and safety officer must identify the training inadequacies and make the necessary recommendations to the training staff. A monitoring process needs to take place to ensure that the training program is enacted.

HEALTH AND SAFETY OFFICER'S RESPONSIBILITY

The program manager of the safety and health program has duties and responsibilities assigned as part of the post-incident evaluation and analysis. This is an important function, based upon the fact that negative aspects of an incident or situation tend to be overlooked or quickly forgotten. The health and safety officer can make a positive impact under these circumstances.

Risk Management

The health and safety officer is the department's risk manager. Most departments spend as little as five percent of their total time at emergency incidents. The remaining time is spent in nonemergency situations. The health and safety officer is responsible for managing these situations. This could include vehicle accidents, accidents or injuries at department facilities, or safety and health issues that need to be addressed. During emergency incidents, the health and safety officer may be involved, or this role may be delegated to either the incident safety officer or the incident commander. The intent is to ensure that the process is covered at each emergency incident.

Investigations

Organizational policy will dictate the investigation responsibilities of the health and safety officer. For nonemergency situations, the health and safety officer has the primary responsibility for conducting investigations based on the nature or severity of the incident. The immediate supervisor may be able to conduct the investigation and then forward the report to the health and safety officer. For example, apparatus or vehicle accidents under nonemergency conditions or personal injuries during daily work functions may require the assistance of the health and safety officer.

Emergency investigations may be conducted by the incident safety officer. The nature and severity of the situation will dictate the involvement of the health and safety officer. If a fatality or serious injury occurs, the health and safety officer becomes the primary investigator.

Training and Education

Based on the investigation reports and analysis, the health and safety officer may determine that a training need exists. This could include training and education in both emergency and nonemergency circumstances. The health and safety officer may conduct the training or ensure that it is conducted by other department staff members.

Standard Operating Procedures

After a review of reports and documentation concerning an incident or after having participated in a post-incident analysis, the health and safety officer must evaluate the affected procedures. The health and safety officer will advise which procedures need to be developed or revised to prevent reoccurrence.

Modifications

The severity of an incident may require the health and safety officer to evaluate the status of personal protective equipment and clothing, apparatus, and facilities. Questions concerning personal protective clothing may include the following.

Was the clothing being properly used?
If a fatality or serious injury occurred, has the clothing or equipment been tested by an independent laboratory?
If clothing failed, why did it fail, and has the problem been discussed with the manufacturer?
Was the proper protective clothing being used for the proper incident (e.g., hazardous materials, infection control)?

Personal protective equipment must be thoroughly examined to determine whether a problem exists. This may require involving the manufacturer or testing laboratory in the process. Does the problem exist due to poor preventive maintenance (e.g., SCBA) or improper specifications (e.g., life safety rope)? The investigation process will help determine what modifications are needed.

The health and safety officer will have to work with department mechanics to determine problems with apparatus. Does the problem exist because of human error (e.g., vehicle accident) or is it due to poor preventive maintenance (e.g., no preventive maintenance program)? Once the problem is identified, the source may be related to training and education as well as procedural development.

Facilities issues may result due to new laws, standards, and regulations (e.g., OSHA Bloodborne Pathogens Standard, Americans with Disabilities Act) or from poor design or construction (e.g., sprinkler system malfunctions). There may need to be an ongoing risk management process to ensure that facilities are upgraded to meet current regulations.

THE OCCUPATIONAL SAFETY AND HEALTH COMMITTEE'S RESPONSIBILITIES

If the department does not utilize a health and safety officer, the occupational safety and health committee may be responsible for managing the safety and health program. The responsibilities may be distributed so that each member or small group of the occupational safety and health committee is assigned a particular function, based on expertise or interest.

Even with a health and safety officer directing the department's safety and health program, the occupational safety and health committee plays a vital role and is a valuable resource to the health and safety officer. The occupational safety and health committee can assist with training, development, or revision of procedures, rewriting of modifications, or any other situation that may require their assistance to complete a project and improve safety and health.

PARADIGM SHIFT FOR HEALTH AND SAFETY

A paradigm shift can be defined as a change or difference in methods of operation. The fire department experiences a paradigm shift through the implementation, development, and use of a solid and effective written safety and health program. This paradigm shift goes even further through the implementation, development, and use of a health and safety officer and an occupational safety and health program. The fire service deserves nothing less.

CONCLUSION

The post-incident analysis is an important part of organizational operations and must be treated as such. Ignoring incident scene inadequacies and mistakes may not have an acute impact on incident operations, but definitely can have chronic effects on the safety and health of personnel. The aim is to critique operations and actions taken, not to point fingers for mistakes made at the incident.

The primary intent is to thoroughly review the incident operations, identifying positive actions and actions that need to be improved or revised. The incident safety officer and/or the health and safety officer must be part of this process. In lieu of these individuals, the occupational safety and health committee may serve or function in these capacities. Input provided by these personnel can identify pitfalls in operations and actions that took place on the incident scene. Some of the recommendations may be quickly completed, others may take time, in terms of writing new procedures, revising or rewriting policy, conducting training, or purchasing new protective clothing and equipment.

The post-incident analysis is a method for improving operations, enhancing the safety and health of personnel, and strengthening the quality of service to customers.

PART 4 | Integration

PROFILE

Chapter 18: Making it Happen

MAJOR GOAL:

To understand how the various components of risk management interact to create the process

KEY POINTS:

- Understand that there are many benefits to effectively managing risks.

- Understand that just as risk management is a systematic process, the approach used when starting the process should be organized as well.

- Understand that effective risk management is a continuous process, not a static event.

- Understand that although Parts I, II, and III of this book address different components of the risk management process, the approach should be seamless when integrated.

- Others have successful risk management programs. Recognize that it is not impossible to have such a program.

- Understand that safety is a value, not a priority.

Emergency Incident Risk Management

PART I
Administration and Organization

- Chapter 1 Overview
- Chapter 2 Intro to Risk Mgmt.
- Chapter 3 Accident, Injury, and Illness Data
- Chapter 4 Laws, Codes, and Standards

PART II
Comprehensive Risk Mgmt. Plan

- Chapter 5 The Mgmt. of Risk
- Chapter 6 Risk I.D.
- Chapter 7 Risk Evaluation
- Chapter 8 Establishing Priorities
- Chapter 9 Risk Control
- Chapter 10 Program Monitoring
- Chapter 11 Training of Personnel

PART III
Emergency Incident Risk Mgmt.

- Chapter 12 Pre-Emergency Risk Mgmt.
- Chapter 13 Principles of Emerg. Incident Risk Mgmt.
- Chapter 14 Incident Safety Officer
- Chapter 15 Personnel Accountability
- Chapter 16 Incident Mgmt. System
- Chapter 17 Post-Incident Analysis

PART IV
Integration

- Chapter 18 Making It Happen

Chapter *18*

Making It Happen

INTRODUCTION

Administrative risk management, nonemergency risk management, pre-emergency risk management, and emergency incident risk management: Why so many terms, and what's the difference between them? Do all risks have to be classified into one of these categories? Relax: terminology is far less important than results.

In the first three parts of this book, we introduced the terminology and attempted to provide definitions and descriptions for the various parts of the risk management process. All the terms are related, however, and the techniques discussed are integrated at the scene of an emergency. Even though time and geography may change and conditions will vary, risks still need to be identified, priorities established, control measures instituted, and follow-up conducted to ensure effectiveness. Sometimes this happens in an office and sometimes at a five-alarm fire. However, the most important message that needs to be transmitted and received is that risk management truly is a *process*, not a static event, and that even though it has many interrelated components, all efforts expended in the risk management endeavor should be directed at achieving positive results.

In this chapter, we shall review those benefits, address how the various components interact to shape the process, suggest some approaches that can be used to initiate the process, and share some tips from those who have experience in managing risk at an emergency incident.

BENEFITS OF EFFECTIVE RISK MANAGEMENT

The benefits of effectively managing risk were discussed in detail in Chapter 2. There are financial benefits, such as savings due to reduction of costs that would have resulted from accidents and injuries. There are less tangible benefits, such as improved efficiency and morale, and a higher degree of efficiency while conducting business.

There is also the benefit that fewer people become injured or ill. The benefit for the organization can and should have a positive impact on the people who are part of the organization. Employees do not leave for work with the intention of getting injured. When an injury occurs, there are severe disruptions and effects on the employee, his or her family and co-workers, and even the employer. The benefits of staying healthy are numerous, and should not be overlooked.

By utilizing the risk management process, risks that confront members are identified, control measures implemented, and the process reviewed to ensure that it remains effective. For many, however, the challenge is deciding how and where to start this process.

A good place to start is with efforts that may already be underway. Many believe that there are no such current efforts in their organization, but that is rarely true. SOPs are in place, there is some semblance of a written safety program, a safety committee exists, or a health and safety officer has been appointed. It is very rare to find an emergency services organization that has to start from scratch.

Just as the risk management process is an organized approach for identifying and controlling risks, the approach used when starting the process should be organized as well. Basically, the following steps provide a good way to start.

> Review this book. It does not have to be repeatedly read from cover to cover. When used as a reference, it will provide answers to many of the most commonly asked questions.
> Inventory the existing efforts. These may include SOPs, safety programs, compliance efforts, protective clothing and equipment purchase and replacement policies, incident management and personnel accountability systems, and any other factors that may influence the health and safety of members.
> Based on the review of the existing efforts, identify what needs to be done.
> Establish and execute a game plan for doing what needs to be done.
> Review the efforts, to ensure that appropriate actions are being taken.

For example, a fire and EMS department wants to establish a risk management program. In a typical year, several employee injuries are reported. Back injuries predominate, but a few slips and falls, cuts and bruises, and an occasional unprotected exposure to a communicable disease also occur.

Making It Happen 281

The recently appointed health and safety officer starts by reviewing member injury statistics for the past three years. There turn out to be no surprises there. Next, SOPs are reviewed to ensure that they appropriately address health and safety, based upon the analysis of the accident and injury statistics. Armed with this information, the decision is made to beef up the SOPs dealing with patient handling, and to create new SOPs to better address communicable-disease exposure control issues. Over time, the health and safety officer will attempt to identify potential risks to the organization and its members, but the initial efforts will help to reduce both the frequency and severity of the injuries and exposures that are already occurring.

Should the process stop once the actual loss areas have been addressed? Definitely not! Effective risk management requires a continuous, consistent review of what has been, what may be, and what can be done to positively impact both of those.

PROCESS VERSUS EVENT

The safety and welfare of fire service, EMS, or industrial brigade personnel operating under emergency and nonemergency conditions is one of the principal concerns of an organization. These personnel are not expendable and cannot be easily replaced. Each organization invests time, money, and effort into training and educating personnel to become efficient, productive, and resourceful. The bottom line is the organization, and its members must take the necessary steps to provide for a safe and healthy work environment.

Through the use of the various types of risk management—administrative, nonemergency, pre-emergency, and emergency—an organization can establish an effective program to combat the occupational accidents, injuries, and illnesses, reduce the exorbitant workers' compensation costs, and combat the apathy that produces resistance to the metamorphosis to risk management and safety, an apathy that continues to plague emergency service organizations.

This is an ongoing process rather than a one-time event. As the safety and health program develops, as laws, codes, and standards change, as accident and injury data are analyzed, as department operations are modified, and as new programs are introduced, the risk management program must be continuously updated to reflect these modifications. Remember, do not get discouraged if control measures do not work; revise them and try again.

SAFETY IS A VALUE

In any organization, safety must become a value, not simply a priority. This ensures that safety becomes a mandatory part of all standard operating procedures and operations of the organization. Safety is not something that is sought "when there is time for it" or "when we get around to it." Tradition has been a very large part of the fire service, yet continues to hinder the development and growth of the safety and health process.

Throughout this book, many procedures have been described that will reduce the risk associated with emergency scene operations. Hopefully, we have provided the necessary tools that will enable an organization to develop, implement, and operate an effective risk management program. All the components are integrated into a final product.

TIPS FOR MAKING IT HAPPEN

In order to develop a successful risk management program, there are several basic elements that need to be in place. These elements will assist in understanding the process, writing the risk management plan, and implementing the procedures in the organization. The basic elements are as follows:

Thoroughly understand the classic risk management model.

Work closely with your organization's risk manager and ask for guidance when needed.

Educate the risk manager on firefighter or responder safety and health issues and how they will serve to enhance the risk management plan.

Solicit and maintain support for the risk management program from the fire chief or top administrator.

Canvass comments from department officers and members during the development of the process.

Organizational accident and injury data will serve as a valuable resource for the development of the risk management plan.

Do not get discouraged, and do not give up!

SAMPLE RISK MANAGEMENT PLAN

A sample risk management plan is included to assist with the development of a risk management plan for your organization. It is to be used in any manner that will help in developing and implementing such a plan.

SAMPLE RISK MANAGEMENT PLAN

IDENTIFICATION	FREQUENCY	SEVERITY	PRIORITY	ACTION REQUIRED / ONGOING	CONTROL MEASURES

Figure 18.2. A blank Sample Risk Management Plan form appropriate for photocopying and use.

THE FUTURE

The risk management process will continue to be integrated into the daily operations of emergency service organizations. As our customer service roles continually expand, risk management must be blended into these operations. As budget constraints and financial adversity continue to affect operations, staffing, equipment purchase, and other necessary components needed to maintain adequate customer service, the financial aspect of risk management becomes more and more important.

Organizations cannot remain stagnant and become comfortable with the current method of doing business. The fire service has had the opportunity to experience change with the advent of NFPA 1500, *Standard on Fire Department Occupational Safety and Health Program.* Some organizations have taken advantage of this program, but others have not. Currently, there is the option to comply with the standard rather than to be mandated to do so.

Due to the poor safety and health record experienced by the fire service, eventually there may be mandatory regulations to comply with safety and health standards. The mandates may be enforced by the Occupational Safety and Health Administration, insurance companies, and/or the authority having jurisdiction. When the financial burden becomes too great, methods will be instituted to control costs. The mandates will not be one-dimensional but, rather, all-encompassing.

In conclusion, the process offers a method to develop and implement a proactive program that will enhance all operations. The primary component of this process will be to reinforce and strengthen the commitment to safety and health and ensure that safety is an organizational value, not simply a priority.

Appendix A

Common Risks and General Control Measures

Control measures against:

I. Risks to members
   A. Strains/sprains
      1. Training and education
      2. Ergonomic task analysis and redesign, where practical
      3. Physical conditioning
      4. Use of appropriate equipment for task
   B. Cuts and bruises
      1. Use of appropriate personal protective equipment
      2. Awareness training
   C. Stress
      1. Training and education on risk factors and appropriate controls
      2. Medical evaluations (pre-placement and follow-up)
      3. Physical conditioning
      4. EAP program access
   D. Health exposures
      1. Training and education
      2. Vaccinations and immunizations
      3. Use of appropriate PPE for body substance isolation
   E. Falls
      1. Training and education
      2. Issuance and use of appropriate footwear
      3. Maintenance of floors, stairs, ladders, etc., that are under the control of the department
      4. Proper "housekeeping"

Emergency Incident Risk Management

PART I
Administration and Organization

- Chapter 1 — Overview
- Chapter 2 — Intro to Risk Mgmt.
- Chapter 3 — Accident, Injury, and Illness Data
- Chapter 4 — Laws, Codes, and Standards

PART II
Comprehensive Risk Mgmt. Plan

- Chapter 5 — The Mgmt. of Risk
- Chapter 6 — Risk I.D.
- Chapter 7 — Risk Evaluation
- Chapter 8 — Establishing Priorities
- Chapter 9 — Risk Control
- Chapter 10 — Program Monitoring
- Chapter 11 — Training of Personnel

PART III
Emergency Incident Risk Mgmt.

- Chapter 12 — Pre-Emergency Risk Mgmt.
- Chapter 13 — Principles of Emerg. Incident Risk Mgmt.
- Chapter 14 — Incident Safety Officer
- Chapter 15 — Personnel Accountability
- Chapter 16 — Incident Mgmt. System
- Chapter 17 — Post-Incident Analysis

PART IV
Integration

- Chapter 18 — Making It Happen
- Appendix A — Common Risks and Gen. Control Measures

Appendix A

      5. Use of appropriate antifall equipment, when appropriate
      6. Appropriate use of antislip measures (e.g., sand on ice)
  F. Exposure to fire products
      1. Training and education
      2. Issuance of mandatory SCBA policy
      3. Issuance of SCBA facepiece seal SOP
      4. Use of appropriate PPE, including PASS devices
  G. Vehicle-related risks
      1. Training and education
      2. Use of personal restraint devices
      3. Appropriate storage and securing of equipment
      4. Use of high-visibility garments when operating on or near roadways
  H. Terrorism/violence
      1. Training and education
      2. Interagency agreements
      3. Use of appropriate PPE
  I. Environmental
      1. Training and education
      2. Use of appropriate personal protective equipment
      3. Appropriate rehab facilities/personnel available
      4. Issuance of SOPs appropriate for local conditions
  J. Incident-related risks
      1. Utilization of incident management system
      2. Utilization of personnel accountability system
      3. Use of appropriate PPE, including SCBA
  K. Other risks
      1. Awareness training and education
      2. Periodic revision of process

II. Risks to apparatus
  A. Vehicle collisions
      1. Training and education
      2. Intersection control devices
      3. Operator licensing
      4. Maintenance and appropriate use of emergency warning devices
      5. Response policy
  B. Physical damage
      1. Public education
      2. Maintenance and appropriate use of emergency warning devices
  C. Theft
      1. Training and education
      2. Adoption and use of appropriate security procedures

D. Vandalism
      1. Training and education
      2. Adoption and use of appropriate security procedures
   E. Age (wear and tear)
      1. Routine, preventive maintenance
      2. Apparatus replacement program
III. Risks to equipment
   A. Theft
      1. Use of adequate security procedures
         a. In facilities
         b. For apparatus, both within and outside the station
      2. Positive identification on all equipment
      3. Maintenance of appropriate records of serial numbers, etc.
      4. Effective communication
         a. With law enforcement authorities
         b. With the public
         c. With legal counsel
   B. Loss
      1. Training and education
      2. Positive identification on all equipment
      3. Positive latching devices, where appropriate
      4. Appropriate organization and recordkeeping ("a place for everything, and everything in its place")
      5. Periodic inventories
   C. Deterioration (wear and tear)
      1. Training and education on proper use and care
      2. Performance of recommended preventive maintenance
      3. Maintenance conducted by certified personnel, where appropriate
      4. Periodic inspection
      5. Equipment replacement policy
   D. Obsolescence
      1. Equipment replacement policy
      2. Conduct appropriate pre-purchase research
IV. Risks to facilities and property
   A. Fire
      1. Observance of appropriate fire prevention activities
      2. Installation of fire suppression systems
      3. Installation of fire detection and notification systems
   B. Other damage (wind, water, etc.)
      1. Routine, preventive maintenance
      2. Special precautions for anticipated events (storms, floods, etc.)

Appendix A

   3. Engineering analysis and recommendations, if indicated (roof snow load limits; earthquake preparedness, etc.)
   C. Age (wear and tear)
      1. Routine, preventive maintenance
      2. Capital improvements program
   D. Vandalism
      1. Adoption and use of appropriate security procedures
      2. Public education
V. Financial Risks
   A. General risks
      1. Purchase of all necessary insurance coverages
      2. Utilization of appropriate accounting policies and procedures
      3. Utilization of appropriate security for agency assets
      4. Effective purchasing policies and procedures
      5. Financial checks and balances, so control is maintained
   B. Lawsuits
      1. Training and education
      2. Knowledge and observance of all applicable laws, codes, and standards
      3. Effective communication
         a. With legal counsel
         b. With policy makers
         c. With customers
         d. With the media
   C. Loss of customer confidence (and funding)
      1. Effective communication (see above)
      2. Public education
      3. Adoption of and adherence to customer relations program/policy
      4. Awareness training and education for members
   D. Mismanagement
      1. Training, education, and, where appropriate, certification
      2. Use of periodic reviews/audits by someone outside the organization
      3. Use of organization-wide performance appraisal system
      4. Unanticipated risks
         a. Training and education
         b. Use of periodic reviews/audits by someone outside the organization

Appendix B

Virginia Beach Fire Department Risk Management Plan

Purpose

The Virginia Beach Fire Department has developed and implemented a Risk Management Plan. The intent of this plan is:

To limit the exposure of the Fire Department to situations and occurrences that could have harmful or undesirable consequences to the Department or its members

To provide the safest possible work environment for the members of the Fire Department, recognizing the inherent risks of the Fire Department's mission

Scope

The Risk Management Plan is intended to comply with the requirements of NFPA 1500, *Standard on Fire Department Occupational Safety and Health Program*, specifically paragraphs:

2-2.1 The fire department shall adopt an official written risk management plan that addresses all fire department policies and procedures.

2-2.2 The risk management plan shall cover administration, facilities, training, vehicle operations, protective clothing and equipment, operations at emergency incidents, operations at non-emergency incidents, and other related activities.

2-2.3 The risk management plan shall include at least the following components:

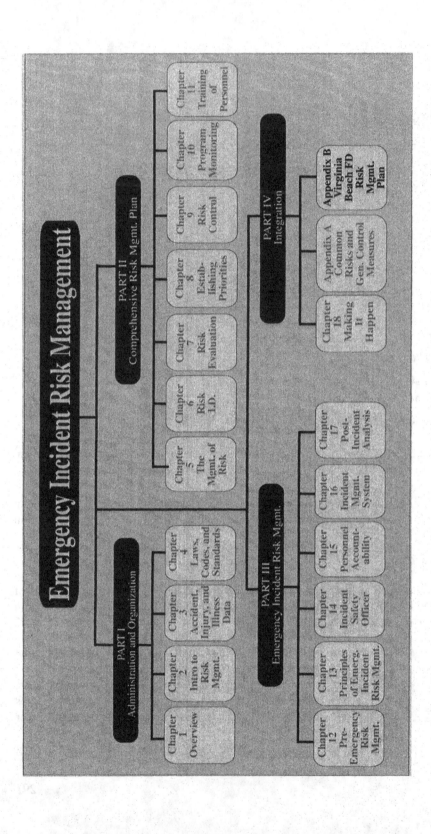

Appendix B 293

Risk Identification: Potential problems
Risk Evaluation: Likelihood of occurrence of a given problem and the severity of its consequences
Risk Control Techniques: Solutions for elimination or mitigation of potential problems; implementation of best solution
Risk Management Monitoring: Evaluation of effectiveness of risk control techniques

Methodology

The Risk Management Plan uses a variety of strategies and approaches to address different objectives. The specific objectives are identified from the following sources of information:

Records and reports on the frequency and severity of accidents and injuries in the Virginia Beach Fire Department
Reports received from the city's insurance carriers and workers' compensation
Specific occurrences that identify the need for risk management
National trends and reports that are applicable to Virginia Beach
Knowledge of the inherent risks that are encountered by fire departments and specific situations that are identified in Virginia Beach
Additional areas identified by Fire Department staff and personnel

Responsibilities

The fire chief has responsibility for the implementation and operation of the department's risk management plan. The department's health and safety officer has the responsibility of developing, managing, and annually revising the risk management plan. The health and safety officer has the assignment of making modifications to the risk management plan, based upon demand and severity of implementing contol measures. All members of the Virginia Beach Fire Department have responsibility for ensuring their own health and safety, based upon the requirements of the risk management plan and the department's safety and health program.

Plan Organization

The risk management plan includes the following:

Identification of the risks that members of the fire department encounter or may be expected to confront, both emergency and nonemergency
Nonemergency risks, including such functions as training, physical fitness,

returning from an emergency incident, routine highway driving, station activities (vehicle maintenance, station maintenance, daily office functions)

Emergency risks, including such functions as fireground activities, non-fire incidents (hazardous-materials incidents), EMS, response to an emergency

Evaluation of the identified risks, based upon the frequency and severity of these risks occurring

An action plan for addressing each of the risks, in order of priority

Selection of a means of controlling the risks

Provisions for monitoring the effectiveness of the controls implemented

Virginia Beach Fire Department Control Measures

Identification	Frequency/Severity	Priority	Summary of Control Measures Ongoing or Action Required	
Strains and Sprains	High/medium	High	1. O 2. O 3. O	Periodic awareness training for all members. Evaluate function areas to determine location and frequency of occurrence. Based upon outcome of evaluation, conduct a task analysis of identified problems.
Cuts and Bruises	Medium/medium	Medium	1. O 2. O	Review SOP on use of PPE, both emergency and non-emergency. Determine if PPE will reduce the number of incidents based upon analysis.
Stress	Low/high	High	1. O 2. O	Continue Health Maintenance Program. Participation in physical fitness program.
Health exposures	Medium/high	High	1. O 2. O 3. A 4. O 5. O	Provide annual retraining on infection control procedures. Re-evaluate city's Health Exposure Control Plan and department's Infection Control SOP. Implement city and department policy and procedures for an occupational exposure to Tuberculosis. Initialize any special training and/or education on an as needed basis. Continue mandatory training and education programs for HAZWOPER.
Falls	Low/low	Low	1. O 2. O	Awareness training for all members to be conducted on a company-level basis. Facility maintenance and housekeeping.
Exposure to fire products	Low/high	Medium	1. A 2. O 3. A 4. A	Re-evaluate department's philosophy on mandatory SCBA usage. Revise department policy and procedures on mandatory usage. Re-training and education of personnel of chronic affects of inhalation of byproducts of combustion. Provide monitoring process of carbon monoxide (CO) levels at fire scenes, especially during overhaul.

Identification	Frequency/Severity	Priority	Summary of Control Measures Ongoing or Action Required	
Vehicle-related incidents	Medium/high	High	1. O	Compliance with department SOPs and state motor vehicle laws relating to emergency response.
			2. O	Conduct/Mandatory department-wide EVOC.
			3. O	Monitor individual member's driving record.
Environmental stress	Low/low	Low	1. O	Revise current department policy relating to "Rehab."
			2. O	Evaluate and implement procedures for "weather extremes."
Terrorism and the workplace	Low/High	Low	1. O	Provide awareness training for all personnel.
			2. O	Develop policy and procedures as indicated by need.
Incident scene safety	Medium/high	High	1. O	Revise and implement department incident management system.
			2. O	Revise current policy on mandatory use of full personal protective equipment, including SCBA.
			3. O	Evaluate effectiveness of the department's personal accountability system and make needed adjustments.
			4. A	Train all officers in NFA *Incident Safety Officer* course.
Apparatus non-collision	Low/high	Medium	1. O	Utilize preventive maintenance program.
			2. O	Properly insure all apparatus and vehicles through risk management.
Equipment loss	Low/medium	Medium	1. O	Review 1994 Accident/Loss statistics and implement loss reduction procedures.
			2. A	Develop procedures for review and recommendation for loss prevention, based upon significant loss ($1,000.00+).
			3. O	Maintain department equipment inventory.
Facilities and property	Low/high	Medium	1. A	Review insurance coverage of contents and facilities for adequate coverage in case of catastrophe.
			2. O	Be certain that all new and renovated facilities incorporate life safety and health designs.
Financial	Low/high	Low	1. O	Maintain liaison with risk management, budget management, and city attorney.

Monitoring Risks: Provisions for Monitoring the Effectiveness of the Controls Implemented

A. The Virginia Beach Fire Department's risk management program will be monitored on a annual basis, starting January 1995, by the health and safety officer.
B. Recommendations and revisions will be made based on the following criteria:

Annual accident and injury data for the preceding year
Significant incidents that have occurred during the past year
Information and suggestions from the division of risk management
Information and suggestions from department staff and personnel

C. Every three years, the risk management program will be evaluated by an independent source. Recommendations will be sent to the fire chief, the health and safety officer, and the occupational safety and health committee.

Appendix **C**

Sources of Additional Information

American Society of Safety Engineers
1800 E. Oakton Street
Des Plaines, IL 60018

Centers for Disease Control and Prevention
1600 Clifton Road, N.E.
Atlanta, GA 30333

Fire Department Safety Officers Association
P.O. Box 149
Ashland, MA 01721

International Association of Fire Chiefs
4025 Fair Ridge Drive
Fairfax, VA 22033

International Association of Fire Fighters
Occupational Safety and Health Department
1750 New York Avenue, N.W.
Washington, DC 20006

Learning Resource Center
National Emergency Training Center
16825 South Seton Avenue
Emmitsburg, MD 21727

Emergency Incident Risk Management

PART I
Administration and Organization

- **Chapter 1** Overview
- **Chapter 2** Intro to Risk Mgmt.
- **Chapter 3** Accident, Injury, and Illness Data
- **Chapter 4** Laws, Codes, and Standards

PART II
Comprehensive Risk Mgmt. Plan

- **Chapter 5** The Mgmt. of Risk
- **Chapter 6** Risk I.D.
- **Chapter 7** Risk Evaluation
- **Chapter 8** Establishing Priorities
- **Chapter 9** Risk Control
- **Chapter 10** Program Monitoring
- **Chapter 11** Training of Personnel

PART III
Emergency Incident Risk Mgmt.

- **Chapter 12** Pre-Emergency Risk Mgmt.
- **Chapter 13** Principles of Emerg. Incident Risk Mgmt.
- **Chapter 14** Incident Safety Officer
- **Chapter 15** Personnel Accountability
- **Chapter 16** Incident Mgmt. System
- **Chapter 17** Post-Incident Analysis

PART IV
Integration

- **Chapter 18** Making It Happen
- **Appendix A** Common Risks and Gen. Control Measures
- **Appendix B** Virginia Beach FD Risk Mgmt. Plan
- **Appendix C** Sources of Additional Assistance

Appendix C

National Fire Academy
16825 South Seton Avenue
Emmitsburg, MD 21727

National Fire Protection Association
Batterymarch Park
P.O. Box 9101
Quincy, MA 02269

National Institute for Occupational Safety and Health
Appalachian Laboratory for Occupational Safety and Health
944 Chestnut Ridge Road
Morgantown, WV 36505

National Safety Council
444 N. Michigan Avenue
Chicago, IL 60611

Occupational Safety and Health Administration
Publication Information (202)219-9631
Public Information (202)219-8151
(or contact your local or regional office)

Public Risk Management Association
1117 N. 19th Street
Suite 900
Arlington, VA 22209

Publications Dissemination, DSDTT
National Institute for Occupational Safety and Health
4676 Columbia Parkway
Cincinnati, OH 45226

Risk and Insurance Management Society, Inc.
655 Third Avenue
New York, NY 10017-5637

United States Fire Administration
16825 South Seton Avenue
Emmitsburg, MD 21727

United States Government Printing Office
Superintendent of Documents
Washington, DC 20402

Select Bibliography

Brunacini, A. V. 1985. *Fire Command.* Quincy, MA: National Fire Protection Association.
Morris, G. P., Brunacini, N., and Whaley, L. April 1994. Fireground Accountability: The Phoenix System. *Fire Engineering.*
NFPA 1500. 1992. *Standard on Fire Department Occupational Safety and Health Program.* Quincy, MA: National Fire Protection Association.
NFPA 1521. 1992. *Standard for Fire Department Safety Officer.* Quincy, MA: National Fire Protection Association.
NFPA 1561. 1990. *Standard on Fire Department Incident Management System.* Quincy, MA: National Fire Protection Association.
National Institute for Occupational Safety and Health. 1994. *Alert, Request for Assistance in Preventing Injuries and Deaths of Fire Fighters,* Publication No. 94-125. Cincinnati OH: U.S. Department of Health and Human Services, Public Health Service, Centers for Disease Control, National Institute for Occupational Safety and Health, DHHS (NIOSH).
Phoenix Fire Department. 1992. Standard Operating Procedure M.P. 206.06. *Incident Critique Sector.* Phoenix, AZ.
Teele, B. W., ed. 1993. *NFPA 1500 Handbook.* Quincy, MA: National Fire Protection Association.
Virginia Beach Fire Department. 1993. Standard Operating Procedure 019. *Incident Command Policy.* Virginia Beach, VA.
Virginia Beach Fire Department. 1993. Standard Operating Procedure 034. *Modified Personnel Accountability Guideline.* Virginia Beach, VA.
Virginia Beach Fire Department. 1993. Standard Operating Procedure P5. *Pre-Incident Planning.* Virginia Beach, VA.

Index

Accident Facts, 77
Accident prevention and training, 141–142
ALS (advanced-life-support) ambulances, 247
Analysis, post incident, 211, 261–274
ANSI (American National Standards Institute), 55
ASTM (American Society of Testing and Materials), 54–55

Bloodborne Pathogens (29 CFR 1910.1030), 144
BLS (Bureau of Labor Statistics), 34
Brunacini, Alan V., 9, 251

CDC (Centers for Disease Control), 50
CFR (crash, fire, rescue) operations, 53
Confidentiality
  medical records, 39
  patient, 39
Consensus standards, 50–55
  American National Standards Institute (ANSI), 55

American Society of Testing and Materials (ASTM), 54–55
National Fire Protection Association (NFPA), 51–54
NFPA 1500 (*Standard on Fire Department Occupational Safety and Health Program*), 51–54
Cost/benefit analysis, 98–100
Costs, indirect, 99
CPR (cardiopulmonary resuscitation), 46

Data
  accidents, injury, and illness, 29–40
  past-loss, 75–76
Data collection/reporting processes, 31–35
  insurance companies, 32
  International Association of Fire Fighters (IAFF), 32
  local jurisdictions, 33
  National Fire Protection Association (NFPA), 33–34
  Occupational Safety and Health Administration (OSHA), 34

305

Data collection/reporting processes (*cont.*)
Public Safety Officers Benefit Program (PSOB), 35
United States Fire Administration, 35

Emergency incident risk management, 177–195. *See also* Risk.
Emergency services, history of health and safety in, 10–11
EMS (emergency medical services), 48
  incidents, 206–207
  injuries, 32
  organization, 137, 139
  representative, 116
EPA (Environmental Protection Agency), 49
Experts, polling the, 19

FAST (firefighters assistance and search teams), 258
Federal mandates, 49–50
Fire insurance, 118
FIRESCOPE (Fire Resources of Southern California Organized for Potential Emergencies), 250–251
Forecasting, 209–211
  emergency medical operations, 210–211
  special operations, 211
  structural fire forecasting, 209–210

GL (general liability), 117–118

Haz mat (hazardous-materials) safety officer, 207
Hazard classification, target, 183
Hazardous Waste Operations and Emergency Response (29 CFR 1910.120), 140
Health and safety officer
  functions, 164–168
  position of, 166
  requiring assistance outside of organization, 168
  responsibility of, 166, 271–273
    investigations, 272
    modifications, 273
    risk management, 272
    standard operating procedures, 272–273
    training and education, 272
Health and safety, paradigm shift for, 274
History, past-loss, 73
HSO (health and safety officer), 25–26, 129

IAFC (International Association of Fire Chiefs), 77
IAFF (International Association of Fire Fighters), 32, 77
Incident analysis, post, 211, 261–274
Incident management system, 245–260
  defined, 146–153
  inception of, 250–251
  incorporating risk management into the, 252–255
  objectives of the, 248
Incident risk assessment, 253–254
Incident risk management, 66–67
Incident risk, toolbox for evaluating, 256–259
Incident safety officer, 164, 197–212
  duties and functions, 203–204
  emergency authority, 202–203
  and forecasting, 209–211
  incident management system, 202
  post-incident analysis, 211
  response criteria, 201

Index

responsibility and authority, 200–201
Incident safety officer, interfacing with the, 270–271
   department procedures, 271
   investigations, 270–271
   training and education, 271
Incident scene monitoring, 204–208
   emergency medical services (EMS) incidents, 206–207
   fireground, 205–206
   hazardous-materials incident, 207–208
   technical rescue, 208
Indirect costs, 99
Information sources, 74–80
   consultants/outside experts, 78
   members of the organization, 76–77
   on-line services, 78
   past-loss data, 75–76
   trade publications, 78–80
   trade/membership organizations, 77
Insurance
   fire, 118
   premiums, 101
   property, 118
   vehicle, 118–119
ISO (incident safety officer), 25

Laws
   codes, and standards
      examination of, 41–57
      influence and effect of, 56
   state, 50
Life Safety Code (NFPA 101), 51
Losses
   actual and potential, 72
   categories, 63–64
      legal liability, 63–64
      personnel loss, 63
      property loss, 63
      time element, 64

Management of risk, 61–68
Management system, incident, 245–260
Mandates, federal, 49–50
MDTs (mobile data terminals), 182–183
Middle management, 22–23
MSDA (material safety data sheets), 47

National Electrical Code (NFPA 70), 51
Net-present-value method, 99
NFIRS (National Fire Incident Reporting System), 35
NFPA 1001, *Standard on Fire Fighter Professional Qualifications*, 139
NFPA 1002, *Standard for Fire Apparatus Driver/Operator Professional Qualifications*, 140
NFPA 1021, *Standard for Fire Officer Professional Qualifications*, 140
NFPA 1403, *Standard on Live Fire Training Evolutions in Structures*, 6, 142
NFPA 1500, *Standard on a Fire Department Occupational Safety and Health Program*, 8–10, 127–128, 144–153
   change with advent of, 284
   data collection and maintenance sections, 33
   post-incident critique, 263
   and proof of equivalency, 138
   utilizing an accountability system, 215
   and written safety procedures, 163
NFPA 1501, *Fire Department Safety Officer*, 25
NFPA 1521, *Standard for Fire De-*

partment *Safety Officer*, 33, 164–165, 202–203
NFPA 1561, *Standard on Fire Department Incident Management System*, 215, 251
NFPA (National Fire Protection Association), 5, 33–34, 51–54, 77
  certifications, 145–146
  developed and issued by, 138
NSC (National Safety Council), 77

Occupational Safety and Health Administration (OSHA), 34
Occupational safety and health committee, responsibilities, 273–274
On-line services, 78
OSHA (Occupational Safety and Health Administration), 44–49
  Bloodborne Pathogens (29 CFR 1910.1030), 48–49
  Hazardous Waste Operations and Emergency Response (29 CFR 1910.120), 45–46
  history, 44
  Industrial Fire Brigades (29 CFR 1910.156), 47
  Permit-Required Confined Space (29 CFR 1910.146), 46–47
  protective clothing, 48
  self-contained breathing apparatus (SCBA), 48
  training and education, 47–48

PASS (personal alert safety systems), 53, 232–233
Past-loss data, 75–76
Past-loss history, 73
Personnel accountability, 213–243
  philosophy, 217
  the players, 233–234
    company officers, 233
    firefighters, 233
    incident commander, 234
    sector officers, 234
Personnel accountability system
  concept of the, 232
  features, 236–243
  positive aspects of the, 219
  reasons for, 217–219
  standard components of the, 234–236
  weaknesses of the, 218–219
Personnel training, 135–156
PIA (post-incident analyses), 128
Planning program, pre-incident, 180–183
Post-incident analysis, 211, 261–274
  benefits and components of the, 264–265
  safety and health issues, 265–270
Pre-emergency risk management, 161–175, 180
  defined, 161
  necessary components for, 161–162
Pre-incident plan, completing the, 183–187
  data sheet, 184
  floor plan, 186
  roof plan, 186–187
  site plan, 184–186
Pre-incident planning program, 180–183
Pre-incident preparation, 187–193
  civil disturbance, 193
  risk management, 193
  tactical worksheet, 187–193
PRIMA (Public Risk Management Association), 77
Priorities, establishing, 93–105
  for action, 103–105
  analysis considerations, 97–103
  analysis factor 1: cost/benefit, 98–99

Index

analysis factor 2: insurance premiums, 101
analysis factor 3: cost, 101
analysis factor 4: ease of implementation, 101–102
analysis factor 5: time required for implementation, 102
analysis factor 6: estimated time for results, 102
analysis factor 7: predicted effectiveness of the control measure, 102–103
balancing the analysis factors, 103
Procedures, revising, 57
Program effectiveness, 125–127
Program evaluation, 128–130
 external evaluators, 130
 internal evaluators, 129
 methodology, 130–132
 recommendations, 133
 results, 132–133
Program monitoring, 123–134
Property insurance, 118
PSOB (Public Safety Officers Benefit Program), 35
Publications, trade, 78–80

Record keeping, reasons for, 35–39
 benchmarking, 38
 data analysis, 36–37
 legal liability, 37–38
 mandates, 37
 medical/insurance, 38–39
RIC (rapid intervention crews), 258
RIMS (Risk and Insurance Management Society), 77
Risk
 assumption, 119
 defined, 15–16, 63, 71
 evaluation, 81–92
 evaluation measures, 83–88
 cost, 86–87

 frequency, 84–85
 impact on the organization, 87
 severity, 85
 time/resources required for rectification, 87–88
 frequency and severity considered together, 88–91
 identification, 69–80
 management of, 61–68
 process of managing, 63–64
 systematic identification of, 72
Risk assessment, incident, 253–254
Risk control, 107–121
Risk control measures categories, 112–116
 accident investigations, 114–115
 the Book, 113–114
 organizational safety efforts, 115–116
 training and education, 112–113
Risk control techniques, 109–119
 avoidance, 111
 fire insurance, 118
 general liability (GL), 117–118
 measures, 111–112
 property insurance, 118
 risk transfer, 116–117
 vehicle insurance, 118–119
 workers' compensation (WC), 117
Risk financing, 119
Risk identification methods
 affected group, 73–74
 line of insurance coverage, 74
 organizational operation/division, 74
Risk identification, recording of findings, 80
Risk management
 administrative, 66–67
 benefits of effective, 280–281
 benefits of, 16–19
 choices, 64–65

Risk management (*cont.*)
  classic model, 67–68
  compliance, 18
  defined, 7–8
  emergency incident, 177–195
  financial benefits, 17
  the future, 284
  goals and objectives, 65–66
  improved efficiency, 17–18
  incident, 66–67
  introduction of, 13–27
  learning lessons, 6–7
  making it happen, 277–284
  overview, 3–12
  pre-emergency, 161–175, 180
  process versus event, 281
  program, 8–10
  program compliance, 24–26
    HSO (health and safety officer), 25–26
    risk manager, 26
  roles and responsibilities, 19–23
    the employees, 23
    middle management, 22–23
    the organization, 19–20
    top management, 20–22
  safety and health, 18
  safety is a value, 281–282
  sample plan, 282–283
  summary of benefits, 18–19
  support and active participation, 23–24
  tips for making it happen, 282
  written plans, 162
Risk management and training, 153–156
  commitment from management, 153
  compliance issues, 154
  forecasting, 154–156
  safety and health program, 154
Risk management training, pre-emergency, 138–140
Risk manager, 26
Risk philosophy, 252–253

Risk retention, 66
Risking
  a lot, 254–255
  a little, 255
  nothing, 255

Safety committee, 115–116
Safety and health issues, 265
  emergency evaluation and analysis, 270
  improving incident safety, 270
Safety and health programs, written, 163–164
Safety officer, incident, 197–212
Safety policy, statement of, 20–21
SARA (Superfund Amendments and Reauthorization Act), 49
SCBA (self-contained breathing apparatus), 9, 11, 48, 53, 233
Scene monitoring, incident, 204–208
SOGs (standard operating guidelines), 113
SOPs (standard operating procedures), 8, 56, 113–114
  and interviews, 132
  manual, 162–163
  modifying, 115
  review of pertinent, 131
  reviewed, 281
  use of, 141
State laws, 50

Time and emergency incident risk management, 67
Toolbox, 168–174
  apparatus and equipment, 172
  components, 168–169
  effective training, 170
  for evaluating incident risk, 256–259
  improving the, 174
  incident management system, 172

Index

personal protective clothing and
    equipment, 170–172
personnel accountability system,
    172–174
standard operating procedures
    (SOPs), 170
TQM (Total Quality Management)
    philosophy, 38
Trade publications, 78–80
Training
    accident prevention and, 141–142
    evolutions in live, 142–144
    mandated, 144–145
    of personnel, 135–156
    pre-emergency risk management,
        138–140
    as a risk control technique, 140–141
    risk avoidance, 140–141
    risk control, 141
    risk transfer, 141
    and risk management, 153–156

USFA (United States Fire Administration), 33

Vehicle insurance, 118–119

WC (workers' compensation), 117
Written risk management plans, 162
Written safety and health programs,
    163–164